Visions of American Agriculture

Other related Iowa State University Press titles

Agricultural Ethics: Research, Teaching, and Public Policy
by Paul B. Thompson. 0-8138-2806-6

Ethics, Animals, and Science
by Kevin Dolan. 0-632-05277-5

Farm Animal Welfare: Social, Bioethical, and Research Issues
by Bernard E. Rollin. 0-8138-2563-6

An Introduction to Veterinary Medical Ethics:
Theory and Cases
by Bernard E. Rollin. 0-8138-1659-9

Planting the Future: Developing an Agriculture that
Sustains Land and Community
by Elizabeth Ann R. Bird, Gordon L. Bultena, and
John C. Gardner. 0-8138-2072-3

Restoring Prairie Wetlands: An Ecological Approach
by Susan M. Galatowitsch and Arnold G. van der Valk.
0-8138-2497-4

Rural Resource Management: Problem Solving for the Long Term
by Sandra E. Miller, Craig W. Shinn, and William R. Bentley.
0-8138-0686-0

Visions of American Agriculture

Edited by William Lockeretz

Iowa State University Press
Ames

William Lockeretz is professor in the School of Nutrition Science and Policy at Tufts University, where he edits the *American Journal of Alternative Agriculture*. His research has covered a broad range of agricultural subjects, including environmentally sound alternative production methods, the interactions between farms and cities, and agricultural energy issues.

First edition, 1997
First paperback edition, 2000

Iowa State University Press
Ames, Iowa 50014
Orders: 1-800-862-6657
Office: 1-515-292-0140
Fax: 1-515-292-3348
Web site: www.isupress.edu

Library of Congress Cataloging-in-Publication Data

Visions of American agriculture / edited by William Lockeretz.
    p.   cm.
    Includes bibliographical references and index.
    ISBN 0-8138-2044-8 (hardcover); 0-8138-2709-4 (paperback)
    1. Agriculture—United States.   I. Lockeretz, William.
S441.V57   1999
306.3'49'0973—dc 21   99-056718

Last digit is the print number: 9 8 7 6 5 4 3 2 1

In memory of Paul and Chloë Sylvain

Contents

Preface

Whenever you touch agriculture, you touch the foundations of society.
—Liberty Hyde Bailey, 1917

This book was written because it is important to think creatively about the future of agriculture. The theme is ambitious, so to do it justice I convened a group of people who could bring imagination and insight to the task.

Why bother?

Within agriculture, it is beyond question that agriculture is fundamental to the national well-being; however, we do not ask those outside agriculture to take this on faith. Indeed, by many common measures—its share of the labor force or gross domestic product (both about 2 percent)—agriculture no longer seems very important. If we don't expect books offering "visions" of other industries of comparable size—automobile manufacturing, for example—why write one about agriculture?

Part of the answer is that the labor force numbers are incomplete. Beyond farming, agriculture is the center of a much larger food system that accounts for about one-sixth of the labor force and GDP. Farmers buy large amounts of fertilizers, tractors, and other production inputs and in turn sell products that are the basis of the food-processing and food-marketing industries.

Also missing from the statistics are people who are strongly tied to agriculture though not actually working in it. In many rural areas, although these are becoming less common, agriculture still is the basis for the local economy and community life. And many people who have left the farms on which they grew up nevertheless remain financially and emotionally involved in agriculture, particularly when the farm stays in the family.

ix

Another way that American agriculture is important for the economy is through exports, which partially offset an unfavorable trade balance. Despite increasing competition, we can expect exports to remain important, thanks to our rich land resources, our highly developed agricultural infrastructure, and the wealth of knowledge held by our farmers.

Finally, agriculture is important—or could be important—in ways that purely economic or demographic statistics cannot capture. Besides commercial exports, the United States contributes international food aid, with significant effects on both the agricultural and nutritional well-being of the receiving countries. As the nation's leading user of land and water, agriculture has significant effects on the environment. Although these effects often are harmful, agriculture could lead the way in showing how productive economic activities can enhance environmental quality. Agriculture once was the stimulus for an entirely new kind of college; revitalizing the land-grant ideal could infuse new vigor into American higher education as a whole. Agriculture also has influenced American ideas about family and community and can contribute positively to the continuing discussions about their future.

In conceiving this project I asked: Will we be offering something different from what others have already written on the future of American agriculture? Discussions of this subject often deal with the near term. They look toward the next farm bill, the omnibus federal legislation that gets revised every four or five years. These efforts are important because the farm bill has a strong short-term influence on agriculture through price supports, funding of research and Extension, and evolution of programs in rural development, conservation, and nutrition.

Other efforts are concerned primarily with changing our production systems. Such work, too, clearly is important. Many agricultural problems, such as environmental degradation and the depopulation of rural areas, arise from the particular ways we raise our crops and livestock.

In contrast, this book takes a longer-range view and is only partly about the areas reached by federal policies. Moreover, it is not about agricultural production as such but rather about the individuals, families, communities (both urban and rural), and institutions that make up agriculture. Although it has become increasingly common to analyze the effects of a new production system beyond the farm, the effects go in the other direction too. Not all problems of agriculture can be solved by starting with our production systems. The kinds of rural communities we have in the future, or the relationships between farms and cities, will affect which production systems get developed and adopted.

As its title makes clear, this book is frankly idealistic—concerned with the agriculture we would like to see, not the one we necessarily expect to see. Yet it also strives to be realistic, setting forth goals that are attainable. We also suggest ways to get there and give many examples of steps already being taken.

By offering "visions," we imply that American agriculture could be better. Actually, the book's underlying question should be: How can American agriculture be made *even* better? We recognize that there is much in American agriculture that should be fixed, but we also recognize that failing to credit what it does right can lead to "solutions" that are much worse than the problems.

Under these general guidelines, each author developed a vision for a particular aspect of American agriculture. To unify these visions, we held two meetings during which we discussed all chapters thoroughly: the first at the start of the project to go over basic ideas, and the second near the end to comment on the final drafts (we had already circulated earlier drafts). As a result, the chapters changed drastically from conception to final version.

Several common themes emerged from this process. One is the importance of looking backward while looking forward. Even in urban, industrialized America, agriculture is strongly bound by tradition and long-held values. Yet American agriculture also takes pride in being dynamic and forward-looking. We must acknowledge the contradiction. Many of the book's authors have built their vision on those traditions and values that can be the foundation of a sound agricultural future, while discarding others that obstruct needed adaptations.

Another theme is the concept of agriculture as more than a supplier of food and fiber. It also provides, or could provide, environmental services, an appealing landscape, a satisfying way of shopping, and a basis for badly needed rural economic development. Moreover, it can nourish not just the bodies but also the minds and spirits of both producers and consumers.

Consequently, as many authors discuss, we need an agriculture that works well when judged not only by the traditional standards of the marketplace but also by its side effects on workers, rural communities, farm family members, and the poor. In turn, this means that the future of agriculture should not be determined just by people who are directly involved with it economically (such as farmers and food processors). Several chapters are concerned with involving other constituencies in shaping our agriculture.

To bring out connections among chapters, the book has been divided into three parts:

- *Values, perceptions, traditions, and cultural expectations:* Some of these support our ideas for improving American agriculture; others need to be modified or discarded.
- *Structural and policy changes:* These are discussed at the level of the individual farm, the community, the region, and the nation.
- *Adaptations to changing conditions:* To ensure this ability, we need a strong system for generating new knowledge, passing it on to those who will use it, and making certain that it is used for socially desirable purposes.

Despite the connections among the chapters, we have deliberately left the book's title in the plural. We do not offer a single, coherent, all-encompassing "vision." That would have required omitting some provocative ideas. Instead, we allowed the pieces to fit together only roughly. Suggesting that to achieve any of these visions requires achieving all of them would only lead to paralysis, leaving none achieved.

We expect the reader to be selective, embracing some visions while rejecting others. (The authors do not necessarily agree with everything their colleagues have written.) No doubt different readers will select differently because we have written for a diverse audience: everyone who, like us, cares deeply about the future of American agriculture. If each reader is inspired by a part of what we offer, we will have achieved our goal.

William Lockeretz

Acknowledgment

We appreciate the valuable contribution made to this book by The Farm Foundation, enabling us to meet at the start of the effort and again near the end.

Part I

1

Past Visions of American Agriculture

David B. Danbom

Throughout our national history we have had two contrasting images of agriculture and the people who practice it. In one, agriculture is a business enterprise, and farmers are people whose significance to the country is primarily economic. In the other, agriculture is a social enterprise—significant primarily for the creation of families, communities, and a nation—and farmers are people whose particular values, traditions, and behaviors make them especially important to society.

Although scholars have noted the apparent contradiction between these images of farming, most people involved with agriculture have been able to hold both simultaneously without causing themselves much difficulty. Consequently, although farmers, politicians, and policymakers have emphasized one of these images when they envisioned agriculture and its future, they have seldom excluded the other entirely. Indeed, they have frequently yoked the two together in their visions, often in a dynamic way. Both are so deeply ingrained in our culture and thinking that both will continue to play a role in our visions of agriculture.

Early Visions

Our contrasting images of American agriculture are rooted in the early colonial period. While some colonists, such as early Virginians and South Carolinians, defined rural America mainly as a place to make money, New Englanders saw it primarily as a place where godly families and communities could be created and maintained. Still others, represented particularly well by the Quakers of Pennsylvania, defined rural America as a good place to do both. The contrasting images of rural America were strengthened in the eighteenth century. American farmers enjoyed enough commercial success to become the most prosperous farmers in the world, but they also were part of a Western Enlightenment culture that celebrated farmers for their natural integrity, goodness, and simplicity.

Until the Revolutionary era, Americans did not dwell very much on their rural and agricultural character or on what that reality might have to do with their identity as a people. That inattention ended abruptly when the prospect and then the realization of independence made the definition and elaboration of a national identity essential to a society that had considered itself English.

In this situation, the characteristics that distinguished the former colonies from the Mother Country and that had historically been regarded as defects when they were examined at all were transformed into indications of superiority. America's ruralness had always set it apart from Great Britain; now it set it above Great Britain. Very quickly, America's agrarian and rural nature became part of the national identity, serving both to distinguish it from Great Britain and to impart a degree of unity to a people fractured on the bases of religion, ethnicity, region, and style of living.

However, the rural and agricultural nature of America served a civic purpose beyond simply providing a component of its identity. The United States was the major republic in a world dominated by monarchies. Republics were notoriously unstable forms of government because they demanded extraordinary self-control and selflessness—or "virtue" in the terms used in the eighteenth century—on the part of the people, who were by nature self-indulgent and selfish. The Founding Fathers, represented most eloquently by Thomas Jefferson, claimed that Americans had the requisite virtue to make republicanism a success in large part because they were mostly farmers. Farmers were virtuous, the agrarians argued, because they lived in natural surroundings, produced basic products all people needed to sustain life, and owned family farms, which made them independent and patriotic. In his most famous formulation on this subject, Jefferson concluded that "those who labor in the earth are the chosen people of God, if ever he had a chosen people, whose breasts he has made his peculiar deposit for substantial and genuine virtue."

Jeffersonian agrarianism almost immediately became an important component of our national identity, and it continues to shape our image of farmers and their image of themselves. The special nature of those who till the soil remains a significant subtext in every debate over agricultural policy. Jefferson and his agrarian contemporaries provided us with an idyllic picture of the countryside, but also an exclusive one. Jefferson's farmers were family-oriented freeholders; invisible in the rural world he sketched were the majority of people, including tenants, farm laborers, indentured servants, and slaves. Women were implicitly there, but in a distinctly secondary position. Jefferson's rural world was a patriarchal one, in which men were "farmers," and women, if they were considered at all, were "farm wives," clearly subordinate to males. Jeffersonian agrarianism, like the republic it meant to buttress, excluded more people than it included.

While Thomas Jefferson is most popularly recognized for his agrarianism, he also was a businessman, an agricultural scientist, and a practical statesman who grasped that agriculture was the key economic enterprise of the new nation. The rest of the founders, even Jefferson's supposed opposite, Alexander Hamilton, also clearly understood the economic significance of farming.

The United States entered the family of independent states as a debtor nation rich in natural resources that had to be exploited and sold on world markets. Its most important natural resource was land, and its main commodities were agricultural. Recognizing this, the founders fashioned a practical agricultural policy for the new nation, even as they celebrated farmers for their noneconomic characteristics.

The object of the first national agricultural policy was to hasten exploitation of land by commercial farmers and facilitate the movement of crops into international trade. In pursuit of these goals, Congress in 1785 passed the Basic Land Ordinance, which delineated an orderly method whereby federal lands could be surveyed and sold to farmers. The rectangular method of survey prescribed by the Basic Land Ordinance incidentally imposed a rigid pattern on the landscape west of the Appalachians. Two years later, Congress passed the Northwest Ordinance, which provided for the systematic extension of government into frontier areas.

Besides converting existing federal lands to private ownership, the government undertook actions to ensure the availability of ample lands for agriculture in the future. The United States systematically extended its borders during these early years, especially with the Louisiana Purchase (1803), which nearly doubled the national domain and which Jefferson believed assured the health of family farming for generations to come. The government also made more agricultural land available by extinguishing Indian claims to desirable areas and removing native inhabitants from them. This policy was so successful that by 1835 few tracts east of the Mississippi River remained in Indian hands.

The United States government and the state governments also took steps to encourage agricultural commerce. Congress chose to keep tariffs low mainly because high duties on imported manufactured products might invite retaliation from our trading partners against our agricultural commodities. Under the Constitution, Congress was responsible for facilitating commerce by improving rivers and harbors, an activity it undertook with enthusiasm, not least because of its pork-barrel aspects. Constitutional scruples limited the internal improvements that Congress was otherwise willing to undertake, but the states were eager to launch projects aimed at supporting commerce. Of these, the most significant was the Erie Canal, completed in 1825, which encouraged settlement of the Great Lakes region by providing for the cheap exportation of its agricultural products.

By this time, the contrasting images of American agriculture had matured and were well elaborated. While politicians, journalists, and commentators on public affairs celebrated the character of farmers, governmental policies were directed almost exclusively at enhancing the agricultural economy. Instead of being contradictory, the images complemented one another; a healthy agricultural economy was a precondition for a thriving rural society, which in turn ensured the success of the republican experiment.

The Civil War and a New Vision

The Civil War was a major divide in American agricultural history. The Union victory destroyed the slave labor system on which Southern commercial agriculture depended and led to a protracted discussion of what should replace it. The former slaves and their supporters favored confiscation of plantations and redistribution of lands to the freedmen as family farms. The behavior of the former slaves who were settled on abandoned lands in Georgia, the South Carolina Sea Islands, and other locales indicated that such a policy would result in an agricultural system aimed more at ensuring family self-sufficiency than at commercial success. Confiscation of lands, although appealing to those seeking independence for the freedmen or vengeance against the South, proved too radical for Congress, which left estates in the hands of owners. Given the choice between creating a small-scale, family-oriented agricultural system and returning, in some form, to a system aimed at commercial production of commodities for export, Congress chose the latter. The result was the emergence of a free labor system in which the former slaves produced commercial crops on the lands of the former masters, mostly as laborers or sharecroppers.

The solution to the question of the future of Southern agriculture was consistent with a national agricultural policy that sought to encourage a commercial and capitalistic farming system, now under an exclusively free labor regime. By the Civil War, observers of American agriculture were concluding that such encouragement required an elaboration and extension of the national agricultural policy.

As the United States industrialized in the nineteenth century, its economy became more complex and sophisticated. New financial, legal, and communications mechanisms revolutionized commerce and banking, and highly mechanized and capitalized manufacturing establishments increasingly displaced the small shops of an earlier era. The new economic person for the new economic era was the professional person, who had esoteric and specialized knowledge gained from a lengthy period of training.

In an environment of qualitative economic dynamism, agriculture remained almost static. New machinery appeared from time to time, such as the reaper for harvesting small grains, but productivity increased mainly because richer lands were opened. Moreover, although scientists were developing an understanding of agricultural processes, especially in soil chemistry, most farmers were heedless of their insights. Amateurs in a world in which the future belonged to professionals, most farmers continued to farm as their ancestors had.

With the laissez-faire Southerners out of Congress, the backwardness of farmers could be addressed, and Republicans did so with two significant measures in 1862. First, Congress elevated the Bureau of Agriculture from a part of the Patent Office to an independent department. The U.S. Department of Agriculture, which went on to achieve cabinet status in 1889, quickly became the main locus of scientific research, economic coordination, and regulation in agriculture. Also in 1862, Congress passed the Morrill Land-Grant College Act, which provided 30,000 acres of federal land to each state for every representative and senator it had, the proceeds from which were to be used to fund colleges that would offer courses in agriculture and the mechanical arts. The Morrill Act arose from a complicated mix of shrewd economic planning, sentimental agrarianism, and pork-barrel politics. Vermont Senator Justin Morrill, the bill's principal author, was genuinely committed to enhancing educational opportunities for farmers' children, in part because he considered them special people. But he and the Republicans who supported this legislation believed that those who took advantage of the opportunities would make agriculture into a stronger segment of the national economy and help it keep pace with industry. Morrill and his colleagues believed that there were scientific principles in agriculture, that farmers should become more thoroughly trained and sophisticated people, and that when they did so, both they and the nation would benefit materially.

The elaboration of the national agricultural policy during the Civil

War reflected a subtle modification of traditional American agrarianism. The founders, while hardly unmindful of agriculture's economic side, believed that farmers benefited the country in large part simply because they were farmers. Producing crops and animals in natural surroundings on their own family farms made them independent, patriotic, moral, and virtuous. The skill with which they farmed was secondary. When Congress passed the Morrill Act and created the USDA, it was suggesting that how farmers farmed also was important. No longer were farmers automatically good; now they were good when they were skilled practitioners attuned to science and technology in the modern age.

To help farmers advance further in the modern age, Congress built on Civil War legislation over the ensuing decades. The Hatch Act of 1887 provided annual appropriations to the states to support experiment stations, attached to the land-grant colleges, that would conduct investigations in agricultural science. And in 1914 Congress passed the Smith-Lever Act, providing matching funds for the creation and maintenance of extension systems to carry scientific developments generated in the land-grant complex out to rural people. The creation of the USDA land-grant complex represents one of the best examples of successful visioning in American agricultural history. It is a major reason that American farmers are among the most technically proficient in the world.

Behind all these legislative measures was the unspoken assumption that most agricultural problems were rooted in the farmers themselves. The implication was that farmers had difficulties because they were unscientific, unsophisticated, unbusinesslike, and generally backward and that the government was there to help. Most farmers disagreed with this position, and the disagreement became vehement in the last quarter of the nineteenth century.

A Farmers' Vision

The late nineteenth century witnessed the increasing commercialization of American agriculture, as farmers raised more crops and animals for market and less for home consumption. Farmers usually participated more fully in the market by choice; improved transportation facilities and a range of increasingly seductive consumer goods were among the magnets drawing them into the markets. Some farmers, however, were forced to participate more fully in commercial agriculture. This was especially true in the South, where subsistence farming was made less viable by rising taxes and fence and trespass laws that excluded farmers and their livestock from private lands they had traditionally exploited.

Whether they were forced to participate in the market or they chose to do so, postbellum farmers discovered that such participation increased their risks. Market prices, over which farmers exerted no control, fluctuated wildly; middlemen often enjoyed monopoly positions; and credit was very expensive when it was available at all. The fact that all prices deflated through most of the thirty-year period following the Civil War was especially stressful for farm debtors.

Farmers responded through a series of protest movements, the most significant of which were the Grange and the Farmers Alliance. These organizations rejected the unspoken assumption of the land-grant colleges and the USDA that farmers were mainly responsible for their own problems and instead blamed agriculture's difficulties on the large-scale forces and institutions that increasingly dominated the American economy.

Agrarian protesters commonly focused most of their attention on the middleman and money problems. The Grange and Farmers Alliance sought to replace middlemen with farmer-owned cooperatives which could manufacture implements, sell consumer goods, insure crops and property, and process and market farm commodities. These cooperatives generally were unsuccessful, mainly because they were undercapitalized, but cooperation would continue to be a popular way for farmers to capture profits that might otherwise go to others. Indeed, cooperatives such as Sun Maid, Ocean Spray, and Land O'Lakes now are massive agribusinesses, and cooperation in general has enjoyed a renaissance among farmers looking to buffer themselves against the economic vicissitudes of the 1980s and 1990s. To curb the abuses of railroads, the Grange favored regulation and the Farmers Alliance wanted public ownership. Agrarian protesters sought to address farmers' money problems with inflation, the Grange through the issuance of greenbacks, and the Farmers Alliance through the monetization of silver. Because the United States was a debtor nation selling commodities on the world market, devising solutions for farmers' marketing problems was more challenging. However, the Sub-Treasury Plan of the Southern Farmers Alliance, which envisioned an institution similar to the Commodity Credit Corporation created in 1933, did include provisions that would have enhanced farmers' control over market timing, if not price.

Agrarian protest reached a climax with the formation of the Populist party in 1892. Contrary to the view of a few scholars on the romantic left, the Populists were not proto-socialist family farmers striving to preserve a moral economy against an increasingly hegemonic industrial capitalist system, though the participation of women and children in their rallies highlighted the fact that farming was a family enterprise. Instead, Populists were commercial producers who participated quite willingly in the late-nineteenth-century world agricultural economy. However, their vision of the economy and how it should function recalled the visions of Thomas Jefferson and Andrew Jackson in the age of John D. Rockefeller and J. P. Morgan. The Populists suggested that the economy's benefits should go mainly to the producers of wealth rather than to the manipulators of markets and money. In their minds, producers had a special legitimacy that derived from their primacy in the economic process. They hardly considered all producers equal—they allied with labor mainly out of political opportunism, and they showed virtually no interest in farm workers and sharecroppers—but their refurbishing of the Jeffersonian vision and their suggestion that agriculture's problems mainly arose beyond the farm gate challenged the assumptions of national agricultural policymakers.

The disappearance of the Populist party after the election of 1896 did not mark the demise of its vision of an agriculture dominated by producers. Indeed, it lived on in the Nonpartisan League, the Farmers Union, the National Farmers Organization, and even the American Agriculture Movement. But the disappearance of the Populists did mean that their vision, and the assumptions behind it, would not fundamentally redirect national agricultural policy.

Many Visions

The greatest crisis for American agriculture and agricultural policy came during the Great Depression of the 1930s. The Depression was so severe that it cast doubt on the whole system of industrial capitalism and encouraged fundamental questioning of its arrangements and assumptions. Many commentators argued that the United States had become too industrialized and was now paying the price. Distributists such as Lewis Mumford suggested partial urban depopulation and resettlement of former city residents on farms where they could mix subsistence agriculture with light manufacturing. In *I'll Take My Stand* in 1931, the Nashville Agrarians, eight academics associated with Vanderbilt University, vigorously upheld the superiority of rural life over an urban existence that they considered excessively materialistic, morally corrupting, and depersonalizing. Social critic Ralph Borsodi was one of several strong supporters of a back-to-the-land movement that sought to create communal farm enterprises for unemployed urbanites. Figures such as Mumford and Borsodi were noteworthy not just for their ideas but also because they were urban social critics who viewed rural America not so much on its own terms as in the context of a national social and economic crisis. As such, they anticipated the generation of "public interest" reformers of agriculture who appeared in the 1960s.

In the crisis of the 1930s, daring alternatives enjoyed respectful attention from policymakers as well as from social critics. Franklin D. Roosevelt's New Deal administration sponsored a Subsistence Homesteads project that drew heavily on the thinking of distributists and back-to-the-landers. The Resettlement Administration (later the Farm Security Administration) experimented with several alternative living arrangements, including communal farms too similar to the Soviet model for conservatives' tastes. Indeed, the depth of the crisis and the sense of possibility it offered were reflected in the fact that the Farm Security Administration and even the Extension Service devoted some attention to such traditional pariahs of the countryside as tenant farmers, sharecroppers, and migratory laborers. With more Americans slipping to the bottom rungs of the economic ladder, those traditionally on the bottom received a bit more sympathy and understanding.

The commercial farmers who dominated most agricultural communities hardly dreamed of a rural society of small subsistence farms or one in which sharecroppers and laborers were respected. But they, too, questioned their place in the economic system and doubted the assumptions and institutions of the national agricultural policy. The idea that agricultural science and the greater productivity it promised were positive was questioned by many farmers, who already produced more than they could sell profitably. Normally conservative producers accepted the Populist argument that farmers were innocent victims of the machinations of bankers, brokers, and middlemen who manipulated the economy to their advantage, and joined organizations that sought to redress perceived injustices, such as the Farmers Holiday Association.

The most important New Deal agricultural policies addressed the concerns of commercial producers by easing some structural handicaps under which they operated. The new programs did not displace the components of the existing national agricultural policy so much as they supplemented them. Government mortgage financing and refinancing programs enhanced the security of farm owners' property. Acreage reduction programs supplemented the incomes of producers of basic commodities, and crop loan programs placed floors under commodity prices, reducing income risk. Because these programs raised domestic prices in a time of suffering, many consumers criticized them. However, they ultimately benefited American consumers by assuring ample supplies at reasonable prices. The effect of the New Deal programs on foreign trade was less benign. Because domestic prices were usually above world market prices, American exports were reduced. The drop in exports was only slightly harmful, however, because our shift during World War I from the world's major debtor to its major creditor meant that sales abroad were no longer essential to our economic health. Finally, Federal Crop Insurance diminished the risks farmers confronted from natural calamities.

Many of these programs were justified on the grounds of saving family farmers, a part of the population that was especially dear to Franklin Roosevelt. This appeal, always potent, was especially powerful in the culturally nationalistic 1930s, when farmers were celebrated as the truest Americans. It is noteworthy, however, that the "family farmers" who benefited were not subsistence producers in Appalachia, sharecroppers in Mississippi, or migrants in California, but commercial commodity producers. Agriculture might be celebrated for social reasons, but its economic problems and potential commanded most of the attention of policymakers.

New Deal programs dramatically increased the role of government in farmers' business decisions. Some farmers also found the government interested in their treatment of the environment. The thirties was a period of increased public awareness of environmental degradation, particularly because of the Great Plains drought and the related Dust Bowl. Government programs removed tens of millions of acres of highly erodible land from agricultural production and introduced techniques to reduce water and wind erosion on tens of millions more. The public interest that the Dust Bowl stimulated in an environmentally benign agriculture waned but never died. By the 1970s, an increasing number of Americans were suggesting that farmers had no right to degrade the natural environment, no matter how special they might be in the agrarian myth.

The fact that there were so many government programs during the thirties, and that they pointed in so many different directions, made it difficult for contemporaries to discern the direction in which agriculture was heading. By the mid-1940s, however, the key programs clearly were those that supported prices and incomes. Programs designed to save the family farm had the unintended effect of lavishing the greatest benefits on the largest producers. Farmers who produced little for the market or who did not produce basic commodities received little benefit, and those on the bottom rungs of rural society, such as sharecroppers and laborers, were excluded almost entirely.

Because the programs diminished the risks confronting commercial commodity producers, they encouraged them to produce as much as they could of commodities that were already in oversupply, frequently filling government storage facilities in the process. The programs also had a harmful effect on the environment because they discouraged diversification and rewarded monoculture. Finally, the programs harmed rural communities. Not only did they encourage farm consolidation and rural depopulation, they devalued the neighborhood and kin cooperation that historically had buffered farmers from risk. In operation, the new national agricultural policy, like that set by Congress during and after the Civil War, was biased in favor of the largest, most sophisticated, and most commercially oriented farmers.

A Vision Is Achieved and Another Proposed

The future to which the fully elaborated national agricultural policy pointed was achieved after World War II. In part because of publicly sponsored agricultural science in the land-grant complex and the USDA, and in part because of the actions of agribusiness corporations, commercial farmers participated in a revolution in productivity.

Part of the revolution involved mechanization. The number of tractors on American farms rose dramatically, rendering mules and work horses curiosities by 1960. Rural electrification allowed further mechanization and dramatically increased productivity in enterprises such as dairy farming and poultry production. The introduction of improved animal breeds and crop varieties, some a generation or more in development, such as the "chicken of tomorrow" and crop varieties tailored for local climatic conditions, ease of harvesting, and other particular characteristics, increased productivity further. Probably most important was the explosion in the number and range of agricultural chemicals. Insecticides such as DDT and chlordane, herbicides such as 2,4-D, fertilizers such as anhydrous ammonia, and a range of antibiotics and growth hormones for animals all became available to farmers in the immediate postwar period. These developments affected every area of agriculture and revolutionized some, such as cotton and poultry production.

Taken together, the components of this revolution dramatically increased the productive efficiency of farmers who took advantage of them and substantially diminished natural risk in agriculture. In tandem with the New Deal programs that had diminished price risks, the productivity revolution helped produce the most risk-free generation of farmers in our history. Risk avoidance was good, but it had unintended negative consequences. Farmers no longer needed to diversify carefully, rotate crops, or cooperate with neighbors to minimize their risks; thus, they imperiled the environment and contributed to community deterioration. Another disadvantage was the fact that as risks diminished, so did rewards. When returns per acre became more steady and predictable, land prices rose. In combination with increasing prices for other factors of production, rising land prices squeezed profit margins. Farmers responded to narrow per-acre margins by increasing the amount of land they farmed, usually by buying the farms of retiring neighbors. The cost-price squeeze and the diminution of risk made expansion prudent, and the productivity revolution made it possible.

By the early 1970s, when agriculture enjoyed a brief period of remarkable prosperity, the visions of many of the architects of the national agricultural policy seemed to have been achieved. Farmers, or at least those who dominated agriculture, were highly capitalized, highly efficient, highly educated, sophisticated producers. They and their families lived in material comfort comparable to that of middle-class urbanites and suburbanites. Living in fully electrified homes, joining country clubs, taking winter cruises, and sending their children to college, many farmers were indistinguishable from other businesspeople. Indeed, some eschewed the title "farmer" entirely, preferring to be called a "grower" or an "agribusinessman."

While pleasing to many observers, this outcome was not universally celebrated. A new generation of socially conscious reformers complained that agriculture's achievements had come at the expense of the environment, the community, and social justice. Latter-day agrarians such as Wendell Berry bemoaned the loss of a sense of family and community and the rise of selfishness and materialism that seemed to accompany the development of modern agriculture. Environmentalists warned of the unsustainability of many agricultural practices, especially the overuse of chemicals. Experts in food safety and nutrition were concerned that abundance was accompanied by a deterioration of quality and healthfulness. And champions of the forgotten people of rural America—small farmers, tenants, and laborers—argued that the rural poor had paid a disproportionate price for productivity increases in modern agriculture.

The deterioration of the agricultural economy in the late 1970s and the agricultural depression of the early 1980s offered the possibility of another major turning point in American agriculture. Angry farmers formed the American Agriculture Movement to protest their economic situation when prices turned downward in the late seventies. The problem of massive farm debt, which was particularly severe for those who had expanded during the land-price bubble of the 1970s, resulted in a good deal of urban sympathy and the formation of urban-based groups such as the Iowa Farm Unity Coalition. While many commercial farmers had defined themselves as businesspeople in the seventies, when they got in trouble they fell back on the tried-and-true agrarian image of farmers as special people with unique social value to the United States. The continuing power of this image was reflected in high public sympathy and in substantial aid from an increasingly urban and suburban Congress.

The agricultural depression led many farmers to reexamine accepted assumptions and reevaluate their work and their lives, just as the Depression of the 1930s stimulated reexamination among their parents and grandparents. Farmers began to question the assumption that scientific progress in agriculture was necessarily positive for them. They were especially skeptical of biotechnology, which they believed would mainly increase surpluses and benefit agribusiness firms, while providing nothing of value to farmers. An acrimonious dispute broke out over bovine growth hormone, which promised to increase milk production. While land-grant scientists and agribusinesses pushed this innovation, small dairy farmers and health-conscious consumer groups opposed it. Some producers also began to question the use of chemicals, on both economic and environmental grounds, and a small but rapidly growing coterie of organic farmers emerged. The depression of the 1980s led many farmers to realize once again the risks inherent in an enterprise they had come to think of as risk free. Farmers developed a new interest in producing for niche markets, in diversifying their crop and animal mix, in bringing more sources of income to the farm family, and in cutting upstream costs and capturing downstream profits through cooperation.

Forces inside and outside agriculture converged to have a major impact on the 1985 Farm Bill. While perpetuating the basic elements of the national agricultural policy, that legislation included such environmentally sensitive provisions as the Conservation Reserve Program, Sodbuster, and Swampbuster. The 1990 Farm Bill built on these beginnings with the National Research Initiative, which included funds earmarked for such areas as sustainable agriculture and rural community development. Clearly, the vision of agriculture that was being introduced in farm legislation differed from that upheld by agricultural science, agribusiness, and many commercial farmers in the quarter century after World War II. Where the earlier vision was mainly economic and stressed productive efficiency, this one was also social and ecological and aimed to create a just and sustainable human and natural environment. And yet, different as they were, both visions were rooted in images of American agriculture deeply embedded in our thought and culture.

Future Visioning

We have written this book because we believe that visions make a difference. They make a difference to individuals by helping them define themselves, commit themselves to larger purposes, and foresee a different, more satisfactory future for themselves and their children. Visions also make a difference to society because they reflect its values and contribute to its identity and because they take on life in public policies. Previous visions of agriculture, when they were rooted in our enduring social and economic images of farmers, have been embodied in institutions and the passage of legislation. When they were not, as in the attempt in the 1930s to create communal farms, they achieved nothing. Sometimes those policies have had the intended consequences, as in the creation of a land-grant complex that eventually had the desired economic effects. Other policies have contradicted the intentions of some visionaries, such as the New Deal policies intended to save the family farm that actually benefited large commercial commodity producers. The 1996 Farm Bill promises farmers the "freedom to farm" without oppressive government regulation. Whether it achieves the desired purpose is not yet clear, but it is based in our image of farmers as economic actors and independent individuals.

When we embarked on this volume, we grappled with the question of why it is important to have visions of agriculture, rather than just strategies, goals, and programs. The answer is that beyond its economic significance, agriculture is important for our culture and national identity. For Jefferson, for Morrill, for the Nashville Agrarians, and for countless others, agriculture is a field of dreams on which we remind ourselves of what we are and project what we may become. Whether any particular vision can be realized in the future will always be in doubt, but that we will continue to consider American agriculture important enough to command our attention and call forth our visions is certain.

2

Agrarian Values: Their Future Place in U.S. Agriculture

Paul B. Thompson

David Danbom shows in Chapter 1 how past visions of American agriculture were built upon many different conceptions of the public good. Divergent strands of agrarian values will give rise to a new vision of agriculture for the twenty-first century. For example, Peggy Barlett explores in Chapter 3 how changing conceptions of the family might be incorporated into a vision of agriculture and rural communities. However, it is not likely that the diverse agrarian values of the past will support a single vision of agriculture for the future. With this in mind, let us start with not one vision, but two.

Two Visions of American Agriculture for the Twenty-First Century

In the first vision, we see the completion of industrial transformation in agriculture, a process that has been underway for several centuries. Food and fiber producers organize their time and resources in response to consumer demands, which themselves have become quite diverse. Some producers operate on a small scale, growing crops or raising animals to supply high-value niche markets, where trends change rapidly and consumers are willing to pay for high quality and timely delivery of unusual products. Others produce large volumes of traditional commodities, perhaps along with new biomass commodities that can be processed with genetically engineered microbes into animal feeds and highly processed human foods. Both types of producer might engage in businesses that were not typical of farming in America's past. For example, our specialized "small" producers might market their products through catalogs that also sell housewares or entertainment services, and our commodity-oriented "large" producers might also operate other enterprises that use their farming skills and equipment, such as producing renewable energy and providing environmental services (Freedgood, Chapter 6).

In this vision of agriculture, communication is largely a process of reading market signals and responding to price changes. This is true both for consumers, who signal their displeasure by spending their money elsewhere, and for producers, who respond to projected demand by deciding whether to use resources for planting tomatoes or stocking up on gingham dresses to sell through the catalog. The rural landscape is not a prominent feature of this vision of industrialized agriculture. Clearly, some commodity-oriented producers still have open fields, but others, especially animal producers, will have production facilities that are more like those of a light manufacturing facility than of a traditional farm. Their "farmstead" will be an assemblage of metal buildings for raising poultry or livestock and for storing feeds and other supplies. Small, high-value production facilities might look very much like the farms of the past, but they might instead be greenhouses or carefully managed hydroponic facilities. Nor would these producers, large or small, feel any need to live on the production premises. Workers and owners alike might live in houses or apartment blocks in nearby communities. They might commute to the farm workplace by car or plane, and they might send instructions to the workforce by phone or E-mail. Night guards will keep the premises secure in their absence.

My second vision of agriculture is more like that of Thomas Jefferson than that of Earl Butz, the former Secretary of Agriculture famed for telling producers to "get big or get out." But it is not really Jeffersonian in the sense described by Danbom. In this vision, we still have a rural landscape populated by farms that are recognizable as such, and people who farm still live in the country. People live on the land and expect to make their living off it. They may have several employees, and they may use various forms of land tenure, but they expect to stay in the same place all their lives, and they make their decisions accordingly. Tomorrow's agrarians will not be simple-minded rustics, however. They will have accepted a variety of regulations that not only restrict pollution but match the scale of operations more closely with ecological processes. They will be more professionalized than they are even today and might be licensed and organized into professional associations. They will accept the need to maintain standards for both the appearance and quality of their products, and many will embrace an educational mission to the broader public by supporting farm vacations for families and on-farm learning experiences for youth.

Many farmers in this vision will be linked to groups of food consumers through community-supported agriculture organizations in which consumers share some of the farmers' risks. Kate Clancy describes in Chapter 4 how community-supported agriculture will vary from place to place and take on a variety of forms. The consumers she envisions do not insist on eating fresh fruits and vegetables out of season. Their participation in the food system will teach them both what to expect from farming and what is truly good food. Better-informed food consumers will take vacations on farms and in farm communities. Teen summer camps will be conducted on farms and ranches where campers will learn the principles of ecological food systems while acquiring the virtue of industriousness as well as a good tan.

That women and minorities will be included in this vision of agriculture is obvious. Barlett (Chapter 3) discusses how their contributions will be appreciated on a more egalitarian basis than in the past. But we may also learn that biology and culture make a difference. If so, we will accept that when many individuals within a gender, race, or ethnic grouping display a special talent or affinity for a certain kind of work, they may do that work in disproportionate numbers. In this vision of agriculture, people talk to one another, and they will need a richer and more meaningful moral vocabulary to reach the shared understandings that make this kind of agriculture possible.

The agriculture of the twenty-first century may resemble *both* of these visions. The industrial vision may dominate in some regions, with the communitarian vision pervasive in others. Yet it seems unlikely that both visions will be realized in the same place at the same time or that both would be equally likely to thrive under a single set of agricultural policies. On one hand, the industrial vision depends on an open market structure that signals producers about what to supply. Furthermore, strict enforcement of grades and standards is necessary for commodity trading, as are the advertising and retailing that permit niche markets to evolve within the food sector. These needs are antithetical to the closed contracting characterized by community-supported agriculture. On the other hand, a communitarian vision requires policies such as zoning or licensing that give the citizens in a region or area more control over what happens there and that permit producers and consumers to form organizations for mutual benefit.

The need for different policies is important for any discussion of values, for values are not simply individual beliefs and aspirations. A vision of agriculture does not simply grow out of what individuals acting on their own decide to do. A vision of agriculture depends on the material conditions of society—the technology, climate, and availability of soil and water—and on public policy. But policy in a democracy is formed through a political process of public debate. Values are crucial to that debate, and it is in debates that agrarian values will shape the future of American agriculture. As advocates of one vision or the other argue for particular policies, they will appeal to the agrarian values of the past. In doing so they will reshape these values in novel ways. Which political values will favor the industrial vision of agriculture, which the communitarian? How will moral discourse shape the agriculture of tomorrow?

Property and Efficiency Arguments

The values of property rights and efficiency have always strongly influenced American agriculture. How will arguments based on these values affect the contest between these two visions of American agriculture? Farmers are among the strongest advocates of property rights in American political debates, and they also have accepted and promoted the norm of using property efficiently. But what do these values mean? Both of these values have both core and distorted meanings. Both can affect public policy in ways that their advocates may not have intended or foreseen.

The Case for Property

When agricultural producers advocate the inviolability of private property, they advocate a principle of noninterference that overrides cost-benefit assessments of public policy goals in all but extreme cases. Property rights protect four dimensions of the owner's interest in land against interference from others. First, the landowner may permit or exclude others' use of the land. Second, income or other benefits derived from land accrue to the landowner, or to a designated beneficiary. Third, the landowner decides which of several competing uses will be made of the land. Finally, the owner may dispose of the land through sale or gift under any terms mutually acceptable to all parties. The moral principle of noninterference means that neither another person nor the state itself may intervene between the landowner and these rights, even when an owner's choice is not consistent with the interests of the majority.

Property law recognizes two principles that constrain the owner's rights. In cases in which a person's use of property violates the noninterference rights of another, the state is obligated to intervene to prevent harm as a function of police power. Police power merely extends the moral principle of noninterference to the protection of others; it therefore is wholly consistent with the moral foundations for strong private property rights. The principle of eminent domain limits the fourth dimension of property ownership by allowing the state to take property (with compensation) without the owner's prior consent, but only when significant public benefits are at stake. In eminent domain, the moral principle of social efficiency overrides the principle of noninterference.

Why should the rest of us accept the belief that farmers should have all four dimensions of noninterference protected? One popular reason is that we all share a desire to have these four interests protected in our own property holdings. Because we have these interests in common, we form an implicit social agreement to protect them. Another popular reason is that these rights actually protect more fundamental liberties, such as the right to life, personal security, and freedom of conscience. Both reasons were more plausible when 85 percent of Americans operated farms rather than today's 2 percent; these private property arguments could be regarded as a vestige of America's past. Yet farmers and many others clearly have a deep attachment to the principles of noninterference. All of us feel a sense of ownership and accomplishment in our work. If another person takes or destroys one's work without a compelling reason, it is natural to feel harmed or wronged. This harm attaches to one's very person, rather than to the thing taken or destroyed. Similarly, one comes to depend on one's possessions in conducting one's daily affairs. Clearly, uncompensated loss of one's possessions compromises the ability to pursue life goals.

All these considerations weigh in favor of strong property rights, but they do not necessarily support the full extent of property rights that farmers might claim. Land uses that emit odors or pollutants may violate the noninterference rights of other people. At most, therefore, farmers' property rights favor them regarding the burden of proof in land-use disputes. Those who wish to curtail farmers' freedom in using their land and other resources must adduce morally compelling reasons for doing so. Where a particular use demonstrably causes environmental harm or poses risks to others, the burden of proof will be easy to meet. Property rights reflect important values that influence our thinking about agriculture, but they are not absolute, and they can be overturned when compelling arguments are offered on the other side.

The Case for Efficiency

Farmers who strive for efficiency are trying to get the best ratio of farming's benefits to its harms and costs. It is obvious why farmers strive for efficiency: efficiency is simply the opposite of waste. Who can argue for waste? Other things equal, it is always good to conserve resources rather than to waste them, and striving to get the most out of your resources is what efficiency means at the core. Yet efficiency has come to stand for some questionable ideas in debates over agriculture. Efficiency demands that the farmer get the greatest production from the resource base. This might mean "avoid waste," but the calculation of efficiency often ignores long-term consequences and almost always leaves out the costs that must be borne by neighbors, food consumers, and future generations.

Partly in response to these problems, agricultural economists have promoted a more comprehensive definition of efficiency. While an individual farmer may make decisions that are efficient for the short run, true efficiency should include long-run consequences and the costs and benefits to all affected parties. Expanding the notion of efficiency tests our belief that industrialization is efficient. When food production industrializes, small farms often are consolidated into larger operations that permit the use of larger technology or that maximize the productivity of labor or management expertise. This is clearly efficient from the perspective of the industrial operator—there is more production per unit of cost—but what about the farmers that lose their farms and must find alternative work? Clearly, they have experienced costs or losses that industrializing producers are not likely to include in their own efficiency calculations. These costs must be included, too. But just as there are additional costs, there also are additional benefits. If the industrial farm produces food more cheaply, consumers will benefit. Because there are more than fifty American food consumers for every farmer, small benefits to consumers quickly accumulate. The aggregate benefits to consumers may more than compensate for the costs borne by farmers who lose their farms. An individual farmer understands efficiency as an admonition against waste, but social efficiency means that we are getting the best ratio of cost and benefit from farming when all affected parties are considered.

While individuals want to pursue their goals efficiently, efficiency as such does little to tell us what our goals should be. We rely on our individual values for setting our goals. But the move to social efficiency requires us to interpret efficiency as a very comprehensive social goal: what is good for the greatest number of people. However, there are at least two problems with this argument. First, it assumes that we know how to evaluate costs and benefits well enough to arrive at reliable measures of efficiency. Some costs and benefits can be measured, but the more usual practice is to assume that whenever people are making voluntary exchanges, both parties benefit. Voluntary trades are assumed to move one ever closer to the most efficient allocation of society's resources, and instead of measuring efficiency, the argument for efficiency is reinterpreted as an argument for allowing all social choices to be made through markets. But of course, the trades one is willing to make depend heavily on whether one is rich or poor, powerful or vulnerable in other respects. This leads to the second problem. Good economists have long recognized that social efficiency depends on the initial distribution of resources and the specific system of property rights. As such, it is hardly a comprehensive moral principle for guiding social policy. Yet not all economists are good, and laziness in our use of language has allowed mindless notions of efficiency to influence our thinking too much.

Property, Efficiency, and Industrial Agriculture

The values of property and efficiency are deeply embedded in the American rural psyche, and they have their place. Arguments that appeal to property or to efficiency make important moral claims that should be respected. Property and efficiency are also important ideas in the argument that articulates what is good about industrialized agriculture and why it might be preferred to the communitarian vision.

First, industrialized agriculture is almost certainly efficient under the current system of property rights. If we allow our social goals to be determined by individual market trades and refuse to reconsider the rules for holding and exchanging property, the evolution toward industrialized agriculture is irresistible. The empirical evidence for this is overwhelming. If the status quo is also the moral ideal, there can be little quarrel with the changes occurring before our eyes.

We must also acknowledge that the industrialized vision will serve consumers' needs well. Consumers who want standardized foods can get them cheaply. Consumers who are willing to pay more for higher quality foods can get them, too. Consumers themselves determine what counts as quality through their market power. If consumers want "organic" or "non-biotech" foods and are willing to pay for them, they will be supplied. If they want foods from family farms or eggs laid by free-range chickens, they will be able to get them, too, provided they pay the extra cost needed to get specialty items to market.

Producers will be able to use their property however they see fit, subject to two constraints. First, if they want to survive in the business climate of the twenty-first century, they will have to use their land and other resources to produce goods that people will buy at a price that covers the cost of production. This is not a new constraint on property use, and there is no reason to think that producers will resist it. Second, they will have to observe the principle of noninterference with the rights of others. On this point, times have changed. People live closer together; production techniques such as applications of pesticides and fertilizers are now known to have potential harmful off-farm effects. Producers have been slow to acknowledge these changes, but their own commitment to principles of noninterference means they must recognize the legitimacy of regulations designed to protect the environment. An industrialized agriculture almost certainly can respond to these regulations with better technology and pass the costs on to consumers.

The environmental responsiveness of industrialized agriculture is important, for some might argue that the communitarian vision is preferable on environmental grounds. The word "industry" connotes smokestacks, toxic waste dumps, noxious fumes, and general environmental degradation, but these effects were associated with chemical industries that arose when environmental effects were unregulated. Whereas industry was associated with pollution, small-scale farms were considered to have either negligible or invisible effects on the environment. But a watershed populated by a thousand small farms with one hundred animals will have the same amount of waste as one populated by ten farms with ten thousand animals, and the small farms will lack the economies of scale to employ the most effective means of containing, processing, and recycling the waste. Arguably, industrialized food producers will better afford the lagoons and recycling technology needed to minimize environmental impact in the future. There is much to be said for industrialized agriculture, and an advocate of the communitarian vision does a disservice if the case for industrialized agriculture is taken lightly.

Stewardship and Self-Reliance Arguments

Property and efficiency were not the only values at work in the agrarian moral discourse of the past. Theodore Roosevelt (no farmer himself) wrote: "If there is one lesson taught by history it is that the permanent greatness of any State must ultimately depend more upon the character of its country population than upon anything else. No growth of cities, no growth of wealth can make up for a loss in either the number or character of the farming population" (cited in McGovern 1967, 28). Roosevelt the warrior might have been calling attention to the fact that military and economic greatness depends on a nation's capacity to produce enough food to support manufacturing and armies, but his emphasis on the word "character" suggests otherwise.

Even a hundred years ago the notion was commonplace that farming (or the right kind of farming) produces character. Country people were thought to possess virtues essential to the conduct of statecraft and democracy. These virtues were not thought to have been produced by formal education, nor by religious devotion, which urbanites shared with rural folk. Farming itself was thought to form the character of country people.

But not just any farming—only the kind that was being done across the American Midwest in the latter half of the nineteenth and early part of the twentieth centuries. Roosevelt and many like him believed that the family-owned and -operated farms that were common during this period reinforced the formation of a virtuous moral character. This sentiment is so foreign to liberal political discourse of the late twentieth century that it requires discussion.

Virtue and Moral Character: In Need of Some Defense

In the aftermath of the civil rights movement of the 1960s and subsequent social movements stressing women's rights and gay liberation, American liberals have developed a moral vocabulary built around claims of right and justice. This vocabulary is sometimes at odds with arguments stressing property rights and efficiency, but all these moral ideas can be used in ways that promote the autonomy of individuals. Autonomy means that an individual can formulate and pursue an individually chosen plan of life, that the individual is not forced to accept life goals or values imposed by others. Clearly, this idea has a deep affinity with the principle of noninterference that supports property rights, although with minority rights and justice for all it also can produce morally compelling arguments for redistributing property more fairly. Even social efficiency can be consistent with individual autonomy, for it does not challenge the notion that individuals are the sovereign judges of what is good or bad for them.

Arguments that stress the formation of character have often seemed opposed to the liberal causes that brought the vocabulary of rights, efficiency, and autonomy to the fore. Arguments based on virtue have suggested that each person has a natural place and role and sometimes are contrary to the aspirations of racial minorities and women seeking new roles. Those who argue for the importance of virtue have celebrated folk tales reinforcing uncontroversial virtues such as telling the truth or keeping promises, but they also defend the social structures of domination that have pushed too many people to the margins of modern societies. They have celebrated the family, church, and ethnic communities that provide the milieu for character-building life experiences but that also seem bent on rolling back the progressive social reforms that liberals have secured since World War II. For their part, social conservatives have been reluctant to scrutinize the arguments that link these institutions to the formation of moral character, preferring to rely on vague threats of moral decay should society fail to heed their warnings.

This means that the language of virtue is in poor condition as the twentieth century nears its end. It is hard to defend a vision of farming that celebrates its salutary effect on moral character today, for the argument is met with justifiable suspicions on every side. "Emotion!" says the scientist. "Reaction!" says the liberal. "Nostalgia!" says the free-marketeer. Even the social conservative is unlikely to welcome an argument that pins so much on farming. Yet I will argue that what is most compelling about the communitarian vision of American agriculture is that we can see how it might promote sound moral character for everyone, not only farmers.

Accounting for Virtue

The virtues are many. Honesty and faithfulness are two associated with moral character. While these virtues are useful for promoting a just society, the tradition of moral philosophy that goes back to Aristotle holds that being virtuous is the ultimate point of morality: one needs a just society because justice is instrumental for virtue, not the other way around. Furthermore, Aristotle thought of the virtues as a mean, as a point of balance among vices. Courage is a virtue, but too much courage becomes the vice of foolhardiness and too little becomes cowardice. The virtues are not absolute, but regulate one another. Society is the ecosystem of virtue. In a just society, the virtues will constrain one another, and the pursuit of morality will not be allowed to spill over into vice.

Aristotle was no agrarian in the usual sense. He thought that those who labored with their hands would forever be deprived of the opportunity for contemplation and learning that was necessary for virtue. By the nineteenth century this aristocratic ideal had faded; the farmer was thought to live in a society that was more conducive to virtue than Aristotle's own. Farmers would need to produce commodities, but the virtue of industriousness would not turn into the vice of greed. Instead, it would be checked by the virtue of stewardship, for farmers could not risk depleting their soils. Farmers would be self-reliant, but this virtue would be prevented from turning into selfishness. The fact that farmers were forced to make a living within a community whose membership and boundaries were fixed by their land also forced them to maintain good relations with their neighbors. So farm communities reinforced the virtues of stewardship and self-reliance and prevented them from becoming obsessions (Thompson 1995).

Any reasonable account of virtue demands that we ask how the material conditions of life reinforce the moral balancing act demanded by virtue. But two points are important here. First, people living under social and material conditions conducive to virtue should indeed tend to be virtuous. This does not mean that they all are virtuous, nor does it mean that they do the right thing all the time. But many of them will possess a virtuous moral character, and they will ask themselves whether acting in one way or another is consistent with sound moral character. Second, even those who do not actually live and work in a virtuous community can come to an understanding of virtue when they see that community as a model. One might argue that we still use the farm communities of the nineteenth century as our models of virtue. For example, they appear repeatedly in the stories we read to our children (Thompson 1993). In other words, a virtuous community produces virtue vicariously, even when it does not actually exist. Its moral importance extends far beyond its borders in space and time.

Stewardship, Self-Reliance, and the Communitarian Vision

The best argument for the communitarian vision described above is that it can serve as a moral lesson for how we all should live. It makes the dependence of human beings on nature more obvious, not only for the farmers who grow the crops and raise the livestock but also for the community members who interact with them in community agricultural programs or educational settings. In these respects it promotes the virtue of stewardship—using nature for human life (rather than preserving it in a museum), but with a fine appreciation of the constraints imposed by the ecosystem of which humans are a part. It helps us discover self-reliance in that we see how we can make a life from this earth. It shows that we must work to make this life, that agriculture involves production, not simply consumption. The principle of self-reliance tells us we must rely on our own initiative. But in the communitarian vision the potential for unrestricted growth of wants is checked by placing self-reliance firmly in a social and ecological context. Clearly, these virtues will not be granted to everyone under the communitarian vision, but the vision itself will create virtue in the breast of many who do not farm at all.

The communitarian vision promotes social solidarity because even self-reliant stewards will experience setbacks and tragedies. In a totally industrial agriculture a financial or political setback might precipitate a decision to relocate, but in the communitarian vision producers are committed to make a living from their land. The fact that land cannot be relocated leads farmers to recognize the mutual self-interest in solidarity. Ironically, it is the communitarian vision that promotes the virtue of industry—simple hard work. On the traditional farm, the link between work and reward is visible even to young children. The consequences of lassitude are learned early, with chores tailored to a child's ability. Yet unlike the urban household in which income is everything, the farm is a place where the performance of chores makes an immediate contribution to family well-being.

In contrast to the industrial vision, it is useful to look at the communitarian treatment of property and efficiency. When considering the industrial vision, we thought of property in terms of rights and of efficiency as a ratio of cost and benefits, as a measure of social welfare. There was little place in this vision for notions of virtue. Yet in the communitarian vision there is also a role for property and efficiency, and both become related to virtue. Property is crucial for self-reliance, but the self-reliant steward is constrained in what may be done with property. One must treat one's own property with respect and with consideration for posterity. Efficiency is a kind of minor virtue of its own, and too little efficiency is the vice of profligacy or waste. Yet in the communitarian vision, we see that single-minded obsession with efficiency makes one penurious and shortsighted. Avoiding waste may be the very essence of good husbandry, but efficiency is still a virtue held in check by all the others.

Now the communitarian vision is assuredly not the agriculture of the nineteenth century, much less the agriculture of Jefferson's Monticello (which was more like Aristotle's than ours). The patriarchal system that dominated the nineteenth century has waned, and we must find room for the social progress that has been made in recent decades. One important difference is that the communitarian vision for the twenty-first century does not imagine 80 percent of our population on the land, at least not full-time. People may spend some time on the land as part of their education for virtue, and farmers may even derive a significant part of their livelihood from providing this service. This means that rural communities must be open to people from diverse walks of life. Another important difference is that farmers will not rely strictly on property rights and God's provenance to maintain the social and material conditions that are crucial to the ecology of virtue. Social policies and community agreements will be needed to ensure that the interdependence between town and country and among neighbors can thrive and that changes in technology do not undercut key social institutions. Making these social institutions into a carefully thought out component of rural life is a change from America's past, but it is perhaps a less dramatic change than those of us who imagine that past from a century's distance might imagine. Creating community has always been work, and it was probably done by women in America's past. The work of building and maintaining community must not be omitted from the tasks we all must undertake in the society of the twenty-first century. Tomorrow's communitarian agriculture will be different from that of the past and better for its commitment to a conception of virtue that is liberated from patterns of domination implicit in agrarian moral discourse of the past.

Conclusion

I have considered two visions of agriculture and sketched the moral arguments that would be used to support each. Since we may find that each vision is realized in different parts of the United States, we need not choose between them. Both visions appeal to important moral ideas, but while there is little danger that property and efficiency will disappear from America's moral vocabulary, the discourse of virtue is not so secure. There is little danger that the communitarian vision of agriculture will overrun the industrial one, but the risks that the reverse will happen are great. Can a communitarian vision prevail anywhere?

I have not offered definitive reasons for preferring the communitarian vision to the industrialized vision. It would have to pass big tests, not the least of which is, "Will it produce enough food?" Yet the question of how much is "enough" depends on having a vision in mind at the outset. Will the Chinese, the Russians, and the residents of Africa and the Middle East continue to need large quantities of American food, or will they, like the Europeans, Southeast Asians, and Latin Americans eventually join the exporters? Will people continue to feed large quantities of grain to animals to maintain diets rich in animal protein, or will health and ecological concerns change the way we eat? Will these questions be answered solely by market forces operating within the existing global distribution of property rights, or will we think deliberately about our needs and consumption patterns, evaluating them in light of how our production activities affect the quality of our life? And of course if the industrial vision prevails in some parts of the United States, it may be more than able to meet bulk food needs.

The answers to such questions depend largely on the difficult and obscure links between vision and values that have been the subject of this chapter. Achieving a vision of tomorrow's agriculture will require arguments. It will require reasons for preferring one vision over another. We should not simply assume that family farms promote environmental quality or that family values arguments are a Trojan horse for those who fear change. Such assumptions undercut our ability to think deeply about our vision of agriculture, and when our own arguments are weak, they cannot be expected to persuade. The arguments of property and efficiency are real arguments that have shaped many social and political transformations of the past hundred years. I have not defeated these arguments here, but perhaps I have suggested how even our claims for noninterference and efficiency might be refocused by thinking again in an ancient way about virtue, about work, and about the natural and social worlds in which we live.

Reference List

McGovern, George S. 1967. *Agricultural Thought in the 20th Century.* Indianapolis: Bobbs-Merrill.

Thompson, Paul B. 1993. Animals in the agrarian ideal. *Journal of Agricultural and Environmental Ethics* 6 (special supplement 1): 36–49.

———. 1995. *The Spirit of the Soil: Agriculture and Environmental Ethics.* London and New York: Routledge Publishing Co.

3

Farm Families in a Changing America

Peggy Barlett

American farm families have many traditional strengths that support a highly productive agricultural system and allow for a satisfying way of life. To take full advantage of the potential contributions of all members, however, farm families will have to explore new behaviors and values that will modify some of their traditions. To carry out the linked tasks of nurturing the farm's plants, animals, and people, the family will have to sustain its uniqueness while building bridges with dispersed communities of kin, neighbors, and allies. The farm family will need support from many institutions if it is to find adaptive responses to the political and economic changes sweeping American society.

The future for farm families will need to look different in four ways:

- Families will have a more flexible division of labor that allows women and men, boys and girls, to enjoy a fairer distribution of work load, acquire a broader range of skills and satisfactions, and contribute more creatively to the family's well-being and the farm's prosperity.
- Families will have more broadly shared patterns of decision making based on trust and cooperation, though recognizing the inherent tension between individuals' needs for autonomy and the farm's need for interdependence between generations. Farm families will develop more flexible options for sharing power and new skills of compromise and negotiation for fostering transitions through life.
- The institutional supports for the farm family will change—from a revitalized Extension Service that expects a less gender-based organization of family enterprise to a fully equitable legal structure of inheritance, credit, taxes, and federal program rights.

I am very grateful to Sonya Salamon and Paul Rosenblatt for helpful advice and generous suggestions on an earlier draft. This chapter has also benefited from the advice of Janet Perry, Wava Haney, Karen Goebel, and Norma Hall.

- Farm families will enjoy greater public awareness of the challenges of agrarian family life and receive enhanced support from media and schools for a cultural diversity that will allow the farm family to thrive.

Before addressing needed changes in each of these areas, I will review some important aspects of farm family life that I expect to endure.

Strengths of the Farm Family

A traditional strength of the farm family is its adaptive flexibility in coordinating daily work and generational reproduction of the farm and family units (Bennett 1982). As parents and children gather around the table for the large noon meal, they exchange progress reports, weigh priorities, and reassess allocations of labor power. An illness within the extended kin group, such as a grandparent in the hospital, leads to the coordination of caregiving among nearby relatives. The harvest mobilizes other ways to pool labor and share equipment. As a child becomes ready to assume farm responsibility, parents may decide to acquire a used tractor from a neighbor to give the child practice in its use, maintenance, and repair. Balancing financial resources also can be part of the farm family's adaptive flexibility: "Household management succeeds in reducing labor, recruitment and supervision costs, increasing incentives for responsible work, sharing risks, and mobilizing the co-residential kin group's potential for education, administration, and cooperative production" (Netting 1989, 230). I see such flexibility as a crucial contribution to the success of the family farm, and it must be sustained in the future.

Another traditional strength of the farm family is its ability to transmit agrarian and entrepreneurial values and a sense of community responsibility. Children learn to accept levels of economic risk and uncertainty that would daunt many a city dweller. They also learn to be creative and innovative in the effort to increase income and productivity. Farm families balance this adventurous, gambling side of farming with a commitment to hard work and daily responsibility for chores. Attention to the welfare of crops and livestock and maintenance of equipment and buildings are part of the craft of farming, in which many take great pride. Another aspect of the agrarian heritage is conservatism and caution in the use of money, especially borrowed money. Many farm families have traditions of frugality in the annual cycle of household and farm expenses (Rosenblatt 1990). I also see farmers' sense of responsibility to living things on the farm as connected to a sense of responsibility to nurture local institutions such as schools, churches, and government. Farm families participate in these institutions more than the average rural resident, and children of farm families more often become leaders in them (Conger and Elder 1994). Children who grow up on farms feel competent and valuable early in life, which seems linked to such patterns of community responsibility and participation. These farm traditions should continue.

The traditional farm family also has a sense of rootedness in time and space. When a family has farmed for many years in one location, its members benefit from long-term knowledge of soils, weather, and the local ecosystem. Farm families have developed complex ways of transmitting this knowledge, along with the farmland and other resources. Successful intergenerational succession has many adaptive aspects. For instance, where large investments must be made, such as in livestock facilities, drainage systems, or fencing, a family that can make multigenerational plans has an advantage. Similarly, young farmers who inherit family land or can buy land from other family members reap significant benefits, such as past soil stewardship and possibly a reduced purchase price. In contrast, it is a great disadvantage when a farmer must seek different fields to rent every few years and has little security of land tenure. Rootedness in one place is also linked to shared spiritual beliefs for many families, and for some, farm work embodies a deep sense of connectedness to the sacred.

Who will be in the farm family of the future? Most U.S. farms now belong to a husband and wife team (the language I use reflects that), but some farms in the future will continue to be operated by other forms of "family" or farm household composition, as they have in the past. For instance, farm families also have long consisted of two brothers or two sisters, a brother-sister team, a unit of parents and unmarried children, a widow(er) and children, and other combinations. I assume that gay and lesbian couples can form farm families, as probably has occurred in the past without public recognition. All the forms of family that exist in towns and cities can also operate farms—the only limitation is the availability of adequate labor for all the tasks needed (some of which can be hired, of course).

Although many aspects of farming have changed greatly in recent years, from huge center-pivot irrigation systems to genetically engineered seeds, many rhythms of planting and harvesting remain the same. So, too, with the farm family. Nursing homes for the elderly and the off-farm employment of farm women do not erase either the daily need to keep family members clean, fed, and healthy or the parental challenge to provide land or employment for succeeding generations. Many cultural patterns that have supported these activities in the past will continue to be useful. However, I also foresee the need for major changes.

Work: Toward a More Flexible and Equitable Division of Labor

A profound transformation in mainstream American society today is the shift in the division of labor by gender, in both the home and the workplace. Farm families, like families all over the country, are reconfiguring daily life and adult expectations. A family that wants to provide all its members with creative opportunities to express their talents and preferences cannot divide the world strictly into men's and women's roles. A rigid, lifelong assignment to certain tasks is less and less useful, especially in a time of smaller families, shorter period of childrearing, longer life spans, more rapid changes in farming strategies, and greater opportunities for men and women to work off the farm. In preserving the adaptive flexibility of the farm family, new ways of opening the traditional division of labor will help to balance the work load, support individual creativity and satisfaction, revalorize caretaking and domestic chores, and reduce some costs of the increasing time pressures on members of farm families.

Balancing the work load is an urgent task for the farm family. Useful in looking at the kinds of work done by farm family members are Rosenfeld's (1985) categories: labor takes place in the domains of the farm, the home, the garden, and the wider community, and in each case can be paid or unpaid. Today, men perform most on-farm work in the United States, both paid and unpaid, and women perform most domestic work and much garden work (unpaid and occasionally paid). Both men and women from farms take jobs off the farm and do volunteer work, though gender-based expectations usually lead women to run the church kitchen while men handle the treasury.

Many parts of this traditional division of labor are shifting, and I foresee that farm families will gain several adaptive advantages by continuing to modify it. Peak work loads, for example, can be shared by more than one skilled person. It is common in some parts of the country, but rare in others, for both husband and wife to be able to operate harvesters or tractors. There is a clear advantage to having several family members available during heavy work seasons. More flexible work assignments also can allow quicker responses to new market opportunities. In recent years, farming has seen some dramatic changes, such as the decline of livestock production on many Midwestern grain farms, which decreases winter work loads, or the rise of labor-intensive vegetable crops, which demands new skills in production and management. Families cannot now foresee exactly what economic or ecological changes are coming, but experience shows that work load flexibility will always be an asset.

A freer expression of talents and interests will also allow farm family members greater personal satisfaction over their lives. One farm husband I interviewed in the 1980s recalled how he had shared doing dishes and giving the children baths when his children were young. He took pleasure in the time with his children and in his contribution to a "fairer" division of the evening's work. In recent years, many nonfarm couples who combine two jobs, family care, and homemaking have found that both women and men find satisfactions in nontraditional tasks. The family greatly benefits from having two cooks and two skilled parents as well as two wage earners. Individuals experience an increased sense of competence and empowerment and avoid some frustrations of assigned tasks that do not suit their temperaments. I foresee farm men and women also enjoying these advantages, although they will have to negotiate tasks more and be willing to compromise.

In these negotiations, both men and women will need support in fostering each other's growth and "making room" for each other's contributions in different spheres. In this regard, I remember the contrast between two Georgia farmers I interviewed. One encouraged his young wife to join him in the fields and taught her how to drive a tractor. The other, according to his wife, laughed at her awkwardness when she first tried to get on the tractor and generally discouraged her from learning about "his" part of the farm work. These men had different notions about what kind of marital partnership they wanted, although both women preferred to be more involved in farm activities and skills. I expect that in the future men and women will discuss these issues more openly and will learn the arts of negotiation so that they can benefit from each other's contributions. Moreover, if children are encouraged to experience a wider range of satisfactions, competence, and rewards, they will be more inclined to stay on the farm and work for its long-term benefit.

A major change in farm life in recent years has been the increasing number of women and men who have taken jobs off the farm. When men are otherwise employed, farming tasks are most often curtailed, but they sometimes fall to women (Gladwin 1991). When women take jobs, men are less likely to help with child and elder care, gardening, or domestic chores. In either event, the family as a whole experiences an expanded burden of work. Women tend to have a disproportionate share, doubling or even tripling their work load. I remember vividly the exhausted face of a Georgia woman at ten o'clock one summer evening. She had worked at a town job for eight hours, had come home and worked four hours more with hogs, and was about to begin to make supper. As she bustled in the kitchen, her husband and children watched television or sat on the porch.

My vision of the farm family of the future includes greater equity in work load. A broader sharing of skills or a redefinition of responsibilities would have required this husband to start dinner or in another way reduce the imbalance of tasks. Hochschild (1989) has studied this "speedup" in middle-class American lives off the farm and notes that, as an annual nationwide average, women do a month of twenty-four–hour days more work than men. This work load exhausts the woman, places stress on her health, and fosters resentment, in turn eroding the caring and intimacy between spouses. Hochschild (1989, 260) concludes that "the most important injury to women who work the double day . . . is that they cannot afford the luxury of unambivalent love for their husbands." The emotional cost of women's resentment is borne by both women and men and is another important reason to move toward greater gender balance in work load in the farm family.

In the future, farm children will be encouraged to learn a broader variety of tasks and to explore a wider range of skills. Men and women will be freer to embrace work that historically has been denied them, thereby sharing more equitably the onerous and unpleasant tasks as well as the community-enhancing and deeply satisfying ones. For this change to be accomplished, farm children must be socialized with a broader range of expectations than is common today. Boys must learn child care, cooking, and basic nutrition of people, not just of hogs and cattle, both at home and at school. Girls, similarly, must learn mechanics and soil science, not just how to put up garden produce and do laundry. Both must learn teamwork, leadership skills, and financial management. This greater flexibility of roles will allow men who delicately care for sows and cows to transcend old cultural rules against using these skills with their parents or children.

My vision of work flexibility does not require that all family members do all tasks; some specialization is more efficient, and I expect that particular members will become the family experts in particular domains. Flexibility of household roles, however, will require selecting lifetime partners with attention to complementarity, since a farm needs a range of skills.

Besides sharing tasks more equitably, the farm family of the future will balance the prestige and value associated with the different domains of farm life. Although researchers disagree about how much women are exploited in the traditional farm division of labor (Adams 1994; Fink 1992; Salamon 1992), the declining number of farm girls who wish to marry a farmer is important evidence. More young men wish to farm than can assemble the land and capital to do so, but girls less often dream of an adult life on the farm. Part of their dissatisfaction concerns the long hours, heavy work load, and greater income fluctuations compared with other occupations, but it also reflects the lower status often accorded to farm women's roles (Barlett 1993). Although ideologies about the desired division of labor on the farm are changing, Hochschild (1989, 211) points to a widespread gap in American society between "the care that a family needs to thrive and the devaluation of the work of caring for it."

There has been much public discussion of the conflicting changes in the value attached to caregiving in several domains: care of elders, time for children, care expressed in community activities and voluntary organizations, and the social contract of loyal service and mutual benefit between employers and employees. Throughout the nation, as we attempt to reformulate what nurture and support mean for us as citizens, individuals, and families, I believe that a new balance in the division of labor, in which both men and women are encouraged to explore dimensions of nurturing, will contribute to a revalorization of caregiving and other aspects of women's traditional work. Women have long experienced the deep satisfactions of caring for those in need, and I expect that farm men will find themselves personally broadened by the opportunity to expand caregiving skills and experience the connectedness and vulnerability of the human condition. In turn, farm men talk about the satisfactions of livestock rearing, healthy crops, and smoothly running machinery, and in the future both girls and boys will be encouraged to learn these forms of caregiving. Although a deep appreciation for the contributions of men and women is already a part of many farm traditions, greater work-task flexibility not only will make the farm more adaptable but will enhance the daily personal satisfaction of those on the farm and their bonds with each other.

Shared Power across Gender and Generations

Paralleling the more fluid division of labor, the farm family of the future will explore a broader distribution of power. Especially among the primary adults responsible for the farm, I foresee that farm decisions will be shared more evenly, reflecting the multiple "stakeholders" involved. This sharing will not stop one partner's voice from carrying more weight because of specialized expertise. However, family farms have often been very male dominated, and some also have been hierarchical by age. Just as nonfarm families in the United States are questioning these imbalances, so too are farm families moving toward a more shared division of power.

Decision making on family farms is particularly complex because it involves three dimensions: assembling the resources over generations to operate a farm, sustaining risky decisions over the farm's annual cycle, and providing for succession to farm heirs. The assembling of the assets, knowledge, and skills needed to farm involves the contributions of many people over many years. It is easy to acknowledge that land and buildings were bought from parents or other kin. Harder to assess are resources such as soil fertility, absence of weed infestations, or disease-free animals. The work that women do to provide healthy food, a clean home, or emotional support also contributes to children's inheritance. Marketing savvy or family political connections can be important contributions. Making decisions concerning annual operations also is complex because farms and ranches are expensive enterprises with low profit margins and high risk. Poor decisions can rapidly undo generations of hard work. In addition, farm children who choose not to stay in rural areas have claims on the value of the enterprise, further complicating both inheritance plans and annual operating decisions.

Some farm families do not have traditions of helping children gain access to land and equipment to begin operations on their own; children are expected to make their own success (Salamon 1985). Given the great cost of buying a farm today and the difficulty of paying such expenses from farm earnings, I predict that more families will borrow traditions that help children to enter farming.

Parents' desires to transmit a farm to their children may conflict with the emphasis of mainstream American society on freedom, independence, and individual rights, an emphasis that usually downplays the lifelong interdependence of spouses and parents and children. Especially in phases of farm development, farm families have to make hard choices about investments and inheritance (Bennett 1982). The reality of multigenerational coordination conflicts with the American ideal of the autonomous individual, free to seek independent satisfaction and achievement. Although American culture also balances autonomy against the values of love, connectedness, and family cohesion, it provides no guidance for farm families seeking to find a successor.

As farm children are more influenced by the nonfarm expectation that each generation will independently choose its own path and create its own career and lifestyle, farm parents will need to be explicit about how these pressures work against the farm's survival. It is true that parents who count on a child's continuing to farm are constraining the child's freedom to choose another occupation. But the parents also are providing community, support, and multiple benefits to the child who stays on the farm. In the future, families will be able to talk more openly about both the benefits and costs of intergenerational coordination. They will seek greater support from nonfarm institutions to find a balance between the desire for independence and autonomy on one hand and the values of continuity, connectedness, and stewardship of family resources on the other.

As men and women share more equally in labor, skills, and status on the farm, I expect that gender will decline as the basis for a greater right to decide about resource use and transmission. Today, depending on family traditions and adult experiences, men's and women's expectations regarding the power balance between spouses can involve separate spheres or a joint farm partnership. Some couples expect the wife to be "the queen of the house" and have full control over child rearing and domestic decisions. Others expect the husband to have the final say in both home and farm decisions. A third orientation, more common among younger couples, sees the spouses in a partnership that allows each a greater authority in one area or another but expects them to share decision making and consultation (Barlett 1993). I expect this third balance of power to become more common. As the family matures, farm women's caregiving often continues and increases in breadth, giving older women a power base as the emotional center of the family. This trend may continue, but if men are more willing to share parenting and affectional bonds, I expect this power base will more often be shared.

Intergenerational relations have been less well studied by social scientists, but as the baby boom generation ages and lives longer than previous generations, we can expect new attention to issues of power, privilege, and skill between adult farm men and women and their elders. Older men's power sometimes used to decline with age, physical weakness, and shrinking information networks. Multigenerational cooperation may become easier on farms than in many other kinds of family businesses, with communication made easier by computers, machinery available to compensate for declining strength, and the farm burdened by increasing paperwork and marketing tasks. Older men and women may thus be able to make valuable contributions for longer into their late years. Extending the parent-child partnership, however, may create tensions with young adults who need to move out of a subordinate relationship with their parents. Some farm family traditions hold sons in a postponed adulthood on their "father's" farm. Daughters-in-law may also long for more independence in decision making. Prolonged junior status can limit family cooperation and harmony as well as the productive contributions of the younger generation. Families must be able to renegotiate authority in several stages over the parent-child partnership so that more balanced contributions and satisfactions for both generations are possible.

Authority over children on the farm is complicated by issues of inheritance as well as daily work load and pay. Many farm traditions expect children to be strictly obedient, but newer desires for more egalitarian relations seem to be gaining ground. Families vary greatly in the extent to which children, both boys and girls, are expected to work around the farm without being paid. Some fathers hire boys as farm hands; others expect them to work free, learning the farm business, until they are brought into a legal partnership. Some girls are expected to serve the family until marriage, but today, more of them often get jobs off the farm. Dilemmas remain: Does a daughter control her whole paycheck or pay room and board? Does a son who works off the farm receive a different inheritance from one who stays on? How does money for education fit into the inheritance equation? In facing these issues, ethnic groups have different traditions (Salamon et al. 1986), and research often has emphasized the conflicts between siblings and generations. I hope that future researchers will study the traditions of cooperation and coordination and will publicize options found useful in the past. Such information will foster a greater openness and flexibility in how such matters can be decided.

Families vary not only in the power shared between parents and children but also in how freely they discuss such matters. Some children can argue for a fair return for their past contributions to the farm estate, but for others, such discussions are taboo. However, the balances of rights and responsibilities may shift; farm families will need broader decision-making skills and may need to develop suitable patterns of negotiation and compromise (Paul Rosenblatt, Department of Family Social Science, University of Minnesota, personal communication). Skills of listening, good communication, and consensus building, while already strong in some farm families, will become increasingly important in the future.

The dance of balance and compromise will continue to be—as it always has been—integral to farm success. Both the long-term survival of a farm and its annual production decisions are affected by considerations of efficiency and profitability and the expendability of family members' preferences, as well as by the loving connections that come from shared work and cooperation over decades. Although shared power can lead to poor business decisions and financial disaster, unshared power can lead to resentment, a breakdown of trust and cooperation, and disaster of a different kind. I foresee farm families shifting the balance of power and learning new skills for more egalitarian decision making in the future.

Toward a Supportive Institutional Context for Farm Families

Governments provide a range of services that greatly enhance the quality of life of farm families, but these services also come with cultural expectations and legal restrictions. Schools, postal services, electric cooperatives, lending institutions, the Cooperative Extension Service, and federal, state, and local funding for hospitals, roads, mental health services, and nursing homes all support the farm family. Some of these services reflect the older division of labor and power, but changes are underway that will allow greater equity for family members.

Tax laws and other legal aspects of farming are becoming more supportive of the farm family. Farms as estates are "asset heavy" and therefore can generate a large inheritance tax liability. Because many farms traditionally were considered owned by the male farmer, with wives not listed on deeds, widows—but not widowers—were vulnerable to big tax bills. Incorporation and other legal measures have reduced this inequity, but some difficulties continue. In the settling of an estate for inheritance purposes or in a divorce, machinery is usually considered owned by the husband. The only way to avoid crediting the full value to the husband's estate, even if the wife was directly responsible for buying it, is careful estate planning.

In the past, divorces in a farm family often reinforced gender inequity by devaluing the woman's contributions to the farm. The woman's unpaid work and monetary contributions were sometimes ignored when the farm was legally owned by the husband or his family. No-fault divorces in many states have now substituted detailed calculations of what each spouse brought to the marriage and what each contributed over the length of the marriage to achieve a "fair and equitable" settlement.

Farm women's independent access to credit has been addressed by national legislation and credit reform, and progress has also been made in women's access to farm program benefits. Federal payments to farmers are an important part of many farm budgets, but historically, women could not apply for and receive program benefits, even if they were the principal farm operator. Federal agencies held that the husband was the principal farmer, and women had to gain access to programs through their husbands. This practice has been adjusted, and a qualified woman farmer can receive payments "as a person in her own right" (Norma Hall, WIFE: Women Involved in Farm Economics, personal communication). One farm woman found, however, that if she accepted farm program benefits, she would jeopardize her social security arrangements. These shifts in federal legislation and regulations are part of a national rearrangement of expectations about family rights and responsibilities.

Fairness across generations also is a difficult issue, and what parents consider equitable may not seem so to children. Karen Koebel of the University of Wisconsin (personal communication) urges farm families to talk over financial and legal aspects of a marriage and to make a written agreement. Especially since first marriages in later years now are more common and there are more second marriages, more spouses today bring developed assets to a marriage. Agreements about assets and inheritance need to be discussed and perhaps renegotiated as time passes and either laws or circumstances change. Successful estate planning is a matter not only for the family but also for the community. Public forums to discuss these private concerns will help the community develop new norms of fairness.

Services delivered to farmers must also recognize the changes underway in the internal organization of the farm family. We need to undo the separation of Extension into an agricultural service oriented toward men and a home economics service whose clients are expected to be women (as John Gerber also suggests in Chapter 12). Farm families who have more shared decision making and work skills will need the support of a gender-neutral Extension Service. The farm family's values and commitments to broader power sharing among various stakeholders—and not just short-term profitability—also must be incorporated into Extension advice.

Greater recognition of women's involvement in farm decision making has already shifted the visibility of women in agencies and on boards of directors, such as the Extension Service, cooperatives, and commodity organizations. Haney and Knowles (1988, 4) have noted that

> farm service providers like agribusiness and the extension service have begun to define the agricultural production unit as including the farm household and thus have started to incorporate farm women into their

promotional and educational programs in new ways. Farm women's new visibility in agriculture seems also to be a factor in the recent efforts to recruit women for technical and university agricultural programs and as agricultural service agents and technical advisors. These changes, in turn, seem to be encouraging more women to farm independently.

Women remain underrepresented, however, in agricultural institutions "that advise, plan, or govern at various levels" (Rosenfeld 1985, 215).

Schools are a crucial institution in supporting the changes I envision for the farm family. Teachers must expect farm boys and girls to seek new skills and satisfactions in the future, and textbooks, clubs, and activities must reflect changing gender roles. Farm families also need support from all levels of education for the difficult trade-offs, discussed above, between individual freedom and multigenerational responsibility and loyalty.

Educational curricula can help by legitimizing farm choices as an alternative to urban American lifestyles and values. Teachers can expect diversity in families' values about cooperation and interdependence versus autonomy and individualism as well as changing patterns of hierarchy and authority. In these conflicts between deeply held values, the farm is only a microcosm of issues increasingly evident in many domains of American society.

Supportive Communities

My vision of the future includes a supportive community, both locally and nationally. The local community, made up interacting kin and neighbors, will expand to include more distant kin and friends. I foresee a supportive national culture that recognizes the value of rural traditions and helps farm families sustain them. Between the local and the national are many levels of "community" that will become more aware of the diversity that farmers bring to American society and will actively value and support it.

An especially serious challenge to agrarian communities is distance. Farm families often have been isolated by difficult weather, poor roads, sparse population, or intensive farm work. Technology greatly enhances the family's ability to stay connected (Sonya Salamon, Department of Human and Community Development, University of Illinois, personal communication). Computers, telephones, and fax machines allow farmers to stay in touch with parents in Florida or Arizona, children in a distant city, political interest groups, and commodity organizations. Especially for the solitary farm woman with young children, whose poignant diaries in the past spoke of great loneliness, the future holds new options for social connection. If rural infrastructure is not allowed to deteriorate (Stauber, Chapter 8), technology will facilitate long-distance mentoring and easier contact with neighbors, church groups, and other peers. As farms get bigger and families smaller and more dispersed, family rituals may shift from the regular Sunday dinner to a monthly or yearly reunion. Family networks can remain strong, however, through these modern forms of communication.

Farm families will continue to foster leadership abilities in their children, but the focus of political activities will have to broaden for the family farm to survive. Farm families are served by vibrant local governments, churches, and schools, and women and men will continue to find satisfaction in contributing to them. But a supportive community will require new coalitions with interest groups beyond the rural area. Farm families will have to be active, as many already are, in shaping the political process to permit the farm to survive when threatened by powerful nonfarm forces. Pressures from corporate farming, suburban and industrial development, and environmental deterioration must all be countered with active grassroots education and political action. "Utopia will require good lawyers, good lobbying, and political alliances with urban folk," insists Paul Rosenblatt (personal communication).

As farm families are increasingly connected to nonfarm groups, they experience pressures toward cultural homogeneity. The distinctive rural orientation of farm life is reduced as farm families watch the same television programs as the rest of the country. Sports and arts events bring greater common ground. Even in architecture, farm homes today look increasingly like suburban brick homes. Such loss of distinctiveness does not by itself challenge farm survival, but these trends are of concern when joined with messages that contradict the agrarian values that support hard choices farm families must make, such as financial frugality, intergenerational planning, and long-term responsibilities to land. Decisions to save rather than spend can mean farm survival in a crisis, but in their interactions with nonfarm institutions, many farm families find the value of frugality belittled. Deeply held religious traditions are one support for alternative values, but even in some rural churches, farmers find themselves outnumbered by members oriented toward more urban lifeways.

Accurate and sympathetic images of farms must appear in all the media to provide a truly supportive community for farm families. Mass media are important socializing agents for children, and they are already a part of the national movement toward gender equity and a redefinition of social roles. The flexible family structure I foresee receives some attention in movies and television today, with scenes of caretaking fathers, career-oriented mothers, and more egalitarian relations among generations. However, farm life itself is rarely portrayed either in mass media or in textbooks, and even more rarely is it portrayed in a way that farmers might recognize. Beyond the stereotypes of "Lassie," media writers, educators, religious leaders, and other shapers of our cultural future will need to fashion some shared understandings of what it means to farm. If farm women and men participate in these institutions and form important constituencies for them, they will help remove old stereotypes.

The frequent discussions of "diversity" in school curricula today are a good step forward but too often reduce "difference" to ethnic festivals and quaint customs. As I have explored here, farm families struggle with hard choices concerning autonomy, family interdependence, farm continuity, traditions of frugality, pressures toward a consumer society, pride in the craft of farming, and other definitions of personal success. These issues vary by region of the United States and by family and ethnic tradition, yet the work and family life of most farm families include aspects that have been—and should remain—distinct from urban American lifeways.

In my vision, there will be greater public knowledge about these differences and increased respect for them. The first step in acknowledging diversity is to learn about it and be able to talk about it, for both those living in the tradition and those on the outside. One example is how rarely our national public discourse includes spiritual language, even though many farmers cherish the spiritual connectedness that permeates their work. "You see God everywhere," say both men and women. Empowering farm families to speak about this issue might also liberate other groups who wish to use such language publicly. Conversation across differences will help farm families find some clarity of their own, for instance, about home making care. Many farm parents share an uneasiness about new habits of commercial daycare for children and food purchases for meals yet are unsure what their uneasiness means. For some, new eating and child-rearing habits feel like a loss, while others celebrate the reduced drudgery and increased creativity in jobs and benefits to children of multiple caretakers. Both those who choose home-cooked meals and homecare for children and those who do not do so need a language to articulate distinctive ways of expressing "care for the family."

These trade-offs demonstrate alternative concepts of ideal daily life, human interrelations, mutual responsibilities, and the meaning of sanctity in daily life. Farm families' challenges in rebalancing family life provide an opportunity for all American families to clarify the changes needed and values to be safeguarded. Farm families deserve broad societal respect and support as they struggle to sustain many important parts of the traditional agrarian life while also searching to integrate new aspects of equity, flexibility, and cooperation.

Reference List

Adams, Jane. 1994. *The Transformation of Rural Life: Southern Illinois, 1890–1990*. Chapel Hill: University of North Carolina Press.

Barlett, Peggy F. 1993. *American Dreams, Rural Realities: Family Farms in Crisis*. Chapel Hill: University of North Carolina Press.

Bennett, John. 1982. *Of Time and the Enterprise: North American Family Farm Management in a Context of Resource Marginality*. Minneapolis: University of Minnesota Press.

Conger, Rand D., and Glen H. Elder, Jr. 1994. *Families in Troubled Times: Adapting to Change in Rural America*. New York: Aldine de Gruyter.

Fink, Deborah. 1992. *Agrarian Women: Wives and Mothers in Rural Nebraska, 1880–1940*. Chapel Hill: University of North Carolina Press.

Gladwin, Christina. 1991. "Multiple Job-Holding among Farm Families and the Increase in Women's Farming." In *Multiple Job-Holding among Farm Families*, edited by M. C. Hallberg, Jill L. Findeis, and Daniel A. Lass, 213–28. Ames: Iowa State University Press.

Haney, Wava G., and Jane B. Knowles, eds. 1988. *Women and Farming: Changing Roles, Changing Structures*. Boulder: Westview Press.

Hochschild, Arlie. 1989. *The Second Shift: Working Parents and the Revolution at Home*. New York: Viking.

Netting, Robert McC. 1989. "Smallholders, Householders, Freeholders: Why the Family Farm Works Well Worldwide." In *The Household Economy: Reconsidering the Domestic Mode of Production*, edited by Richard R. Wilk, 221–44. Boulder: Westview Press.

Rosenblatt, Paul C. 1990. *Farming Is in Our Blood: Farm Families in Economic Crisis*. Ames: Iowa State University Press.

Rosenfeld, Rachel Ann. 1985. *Farm Women: Work, Farm, and Family in the United States*. Chapel Hill: University of North Carolina Press.

Salamon, Sonya. 1985. Ethnic communities and the structure of agriculture. *Rural Sociology* 50 (3): 323–40.

———. 1992. *Prairie Patrimony: Family, Farming, and Community in the Midwest*. Chapel Hill: University of North Carolina Press.

Salamon, Sonya, Kathleen M. Gengenbacher, and Dwight J. Penas. 1986. Family factors affecting the intergenerational succession to farming. *Human Organization* 45 (1): 24–33.

4

Reconnecting Farmers and Citizens in the Food System

Kate Clancy

Good farming means good food; anyone who cares about food has a stake in good farming and in the methods of food production, processing, and distribution that accord with the long-term health and sustainability of farmers, farming communities, and the land upon which they, and we, depend.
—Robert Clark, 1990

Today it is the process of consumption itself that must be engaged, through an act of self-revelation, into taking responsibility for the power it wields without consciousness, and therefore largely without responsibility.
—Daniel Miller, 1995

In the last two decades farmers' markets have sprung up in many places around the country, in posh shopping centers and on inner-city vacant lots. Community-supported agriculture operations, in which consumers pay farmers in advance for a season's worth of fresh produce and sometimes other goods, have grown from a handful to around 550 (Van En 1995). Chefs have started to put the names of their local farmer/suppliers on their menus, and in the summer, supermarket ads carry pictures of local growers.

These are the visible responses to a set of demographic, economic, political, environmental, and social changes that have occurred over the last fifty years. Those changes have weakened the relations between the producers and consumers of food, relations that people all over the country are now attempting to reconstruct. If phenomena like those mentioned above prove widespread and long lasting, it will signal our ability to remake the connections between farmers and consumer/citizens and to fulfill a vision of a sustainable, secure, and democratic food system.

47

That vision is of an agriculture that is much better integrated with all the other components of the food chain. It focuses on relations among all parts of the food system, particularly those between farmers and consumers, that have been neglected in the last half century. It asks both those who produce food and those who eat it to speak to one another's concerns (Clark 1990) and asks all to rediscover or rethink their proper connection to the earth (Berry 1993). Finally, it requires that these new sensibilities be translated into an activism that will assure that the necessary changes occur.

In this chapter I first discuss the major factors that have weakened the links between farmers and the rest of the population. I then describe the components of a food and agricultural system in which these linkages are remade. Finally, I outline the questions of policy and ethics that must be addressed to achieve the vision.

Forces That Have Separated Food Producers and Consumers

When most of the population farmed, the producers and consumers of most foods were the same—family members or close neighbors. As the United States transformed itself into an urban, industrial society, people became more dependent on unknown producers for much of their food supply. At present, many factors contribute to the separation, starting with physical distancing (Kneen 1989). Early in the country's history, perishable and heavy commodities such as dairy and produce were produced closest to a town, while grains were produced farther away; a zone even farther out was used to raise cattle and make the most extensive use of land that was cheaper because it was so far from markets (Cronon 1991). As canals, railroads, and then highways evolved and "unlimited" quantities of oil were discovered and transported to the United States, it became easier to produce any food far from where it would be consumed and to take advantage of climates that allowed year-round production (Goodman and Redclift 1991).

Supermarkets were developed in the 1920s, gradually replacing many small markets. Because more people owned cars, markets were developed on large tracts of land, farther away from central cities and with greater space for merchandising and parking. Processors had more volume to fill in these markets and were ready to capture economies of scale. Over time, the gap between farmers and their ultimate markets was widened to extraordinary distances by several trends: the removal of fruit, vegetable, and meat processing plants from many local areas and centralization in locations closer to the fields and feedlots; the diversification of food companies through direct acquisitions and then leveraged buyouts; and the continued glob-alization of the sources of food. The effect was to make farmers merely producers of raw commodities at the upstream end of the chain and to make even farm families dependent on the larger food system for their sustenance.

These developments made world food markets more reliant on lim-ited fuel reserves; a side effect was to remove the reality of farmers and seasonality from most people's consciousness. To a great extent the concept that nature dictates the foods available in a specific area has been lost, and with it, concern about the loss of farmland and farmers.

Another factor that has alienated farmers from eaters is the chemi-cal "revolution" in agriculture and manufacturing. Pesticide residues on produce and other foods; the belief that damaged soils produce less nutritious food; the use of antibiotics and growth hormones in animal production; and microbiologically caused illnesses in the United States from contaminated beef, poultry, and eggs have undermined the faith that some consumers might have in producers.

The extensive use of additives in food processing has also been sig-nificant in separating the public from farmers. An extraordinary array of additives has allowed processors to add "services" to food and add to their profits; these convenience foods have shifted the share of the food dollar away from the farmer to the processing and retailing sec-tors. Although they are advertised as "healthy" substitutes (the claim is debatable), the development of synthetic dairy and meat foods works against farmers' interests because the ingredient statement often pro-vides no sign that a farmer has been engaged in the production of its contents (often, none has been). The farmer's plight is further exacer-bated by food processors' search for substitute ingredients that will make them less dependent on weather and market cycles (and farmers) and provide them with more stability and control (Goodman et al. 1987).

Processed foods are very heavily advertised, but a small share of ad-vertising dollars is spent on commodities closer to the form that left the farm gate, such as fresh vegetables, fruits, meat, eggs, and milk. This advertising rarely depicts a farm or farmer, nor are farms high-lighted in advertising for heavily processed foods, except as romanti-cized images. Economic concentration has led to a skyrocketing of product introductions and money spent on advertising (Connor et al. 1985) and has made consumer choice much more difficult. The result is a trivialization of food and the further erosion of producer/con-sumer relations.

Although the largest profits by far in the food system are made by processors, and despite predictions that the farm sector will disappear in some areas of the country in the next twenty years (Smith 1992), farmers and ranchers often get blamed for rising food prices. This seems to occur because of the still-honored notion that food should be "cheap" (Browne et al. 1992; DeLind 1992). Accepted now as a tenet of capitalism, cheap food was a phenomenon first pursued by industrialists intent on keeping food prices low so that workers could spend more on manufactured goods (Goodman and Redclift 1991). The goal continues to be a double-edged sword for consumers because one means of ensuring low prices for consumers is to keep wages low (Miller 1995). Various conditions and policies have encouraged overproduction and in most years have kept supplies of raw commodities ahead of demand, so that the farm sector's returns have decreased while food costs have increased. Therefore, the prices paid to farmers are not very closely related to retail food prices, although many consumers do not know this. Similarly, farmers mistakenly think that food is cheap because of their dwindling share of the food dollar. The perception of cheapness allows politicians to keep the minimum wage and welfare benefits low and causes producers and consumers to blame each other for their predicaments.

Food in the United States is cheap when the cost of purchasing food is compared with the average household income, which is among the highest in the world (Browne et al. 1992). But this is misleading. It does not reflect government payments to farmers or for agricultural research. It does not reflect long-term environmental costs that will be borne by future generations. What is more, the large population in poverty lives on far less than the average household income; low-income families may therefore be excused for not believing that food is cheap. Food as it is produced also is not cheap if one includes soil erosion and water pollution, loss of autonomy, job losses that occur as processing becomes more labor extensive and centralized, and declines in food quality and quality standards (DeLind 1992).

These costs have been invisible for several reasons. The first is that the present economic system has paid little attention to them; the second is that social relationships in the food system have almost disappeared to the point that consumers have little reason or opportunity to pay attention to the negative effects the system is having on producers, natural resources, and their communities. Agricultural economists have tended to ignore consumption, defining it simply as people acting in the market solely to secure self-interested needs. Despite the fantasies of some free marketers, consumer food preferences are not always met or taken into account, especially when they arise not out of immediate gratification of wants but out of other concerns such as conservation and the preservation of farms. Yet it is true that people fail to recognize how the actions they take as consumers, such as demanding cheap food, may have adverse consequences on their status as workers (Miller 1995). This inattention apparently spills over to farmers and the environment—a particularly egregious oversight in the case of farmworkers' wages and living conditions (Vaupel, Chapter 9).

In the past several years, as industrialization in the farm sector intensified, food-industry experts intensified their claims that "consumers drive the market." Yet Cronon (1991) and others have observed that the ability of the market to "constrict and obscure relationships" has been expanding for a long time. The sovereignty of "consumer choice" is an important democratic principle, but it becomes dangerous when it masks legitimate concerns with the imperatives, moralities, and responsibilities behind consumption (Miller 1995).

Food policies have been both the cause and effect of many of these problems. Subsidies to farmers, an idea with considerable support at their inception during the Great Depression, became less popular as their goal moved from equity to rewarding bigness. The 1996 Farm Bill, debated in a climate of budget cutting and removal of the safety net for many groups, ordered the end of price supports in seven years. Farmers are less visible in many ways—the population census stopped counting them a few years ago, and most rural development has written off farming as a viable component (Stauber, Chapter 8).

Other areas of government have worked against the maintenance of small and midsized farms; for example, no antitrust actions have been taken to limit the size of enterprises or the economic concentration within a sector. As grain traders and food processors have gotten bigger, farmers of many commodities have had even less influence in setting prices, and power has shifted away from this sector. Food-safety laws and regulations grew enormously as the trust between consumers and producers deteriorated through distance and increased anonymity, and now the expense of complying with many of the laws makes it difficult for smaller operations to stay in business.

A Vision of Connection and Health

These forces have had many positive effects, and some may think that this has been an "innocent progress" (Kramer 1990), but taken as a whole, the present food system is detrimental to the environment, the community, the public's health, and people's ability to accept the natural world and function within it. As opposed to the dysfunctional system described above, the food system I envision is one in which farmers and ranchers in all regions of the country and the world have adequate livelihoods and their unique role in survival and culture is secure. Farmland is preserved for production and aesthetic purposes, food-related employment is increased, and local economies are strengthened. Young people are encouraged and able to choose farming as a career, and communities support that choice.

Because they are essential for the food security of future generations, environmental resources are preserved and protected and food and water supplies become more safe. People become more sensitive to the connection of "soil to sustenance" (Clark 1990) and to their own relationships to and within the natural world. They understand the central role of food in their lives and the relationship of food to health. They recognize themselves as "participants in agriculture" (Berry 1990) and take a greater role in policy making. They recognize the distorting effects that monopolistic practices have on themselves and the culture and challenge the usurpation of polity by economics. They seek to bring their roles as consumers and citizens into better balance and to foster greater equity throughout the system.

The following are some of the major components of a more integrated system:

• *The number of farms and farmers in each region of the country is adequate for the region to be self-reliant at an agreed-upon level relative to the food needs of its population and its food-producing capabilities.* The goal is self-reliance (a state of nondependency and trade built on reciprocity and equity), not self-sufficiency (isolation and full autonomy) (Kneen 1989). It is unreasonable and unnecessary to expect each region to produce all its own food, but all should produce what is possible within various planning and trade parameters.

• *Farmers who diversify production and engage in value-added marketing predominate.* Through direct marketing, subscription farming, cooperatives, and networks, farmers recapture a larger share of the food dollar, build new connections with citizens, and contribute to the viability of rural and urban communities, including low-income areas. Farm workers are recognized for their essential role in food production and are no longer stigmatized and discriminated against. Farmers feel more comfortable in explaining that they see their work as a "calling" and not only a business. "Right-to-farm" claims that protect farmers from nuisance complaints brought by their nonfarmer neighbors are fully accepted, and because of the social, economic, and health consequences of production methods, farmers and communities recognize each other as partners in the generation and adoption of new technologies (Flora 1992).

- *The number of food processors increases and their scale decreases.* There are more local/regional processing and slaughtering plants, and safety and other regulations support and encourage smaller-scale local operations. Processors have more direct relationships with farmers. Craft food processing is in demand, and these activities increase employment.

- *Among many needed rural development policies, some focus on enhancing food-related enterprises.* The number of wholesalers, brokers, distributors, and retailers increases in local areas (both urban and rural), and employment in the sector increases. The total number of food products is lower because of concerns about energy use and health, and consumption of minimally processed foods grows.

- *Institutional and individual food-service operations are committed to "sustaining and strengthening traditions of regional agriculture"* (Culinary Institute of America 1996). They obtain a significant amount of food from local suppliers and educate their customers and clients on the benefits of doing this.

- *Most food production, processing, transportation, and waste use/disposal is conducted in resource-conserving ways that are more decentralized and less energy-intensive.* There is less packaging and more research on and development of smaller-scale machines and processes. The number of large feedlot operations declines, and food-safety and water hazards from animal products are lower because of decentralized production and processing. Sustainable production methods have decreased pesticide and synthetic fertilizer use, so food and water hazards also are reduced.

- *Local governments all have food-system planning bodies.* Urban and rural planners routinely consider food needs and incorporate into planning exercises elements such as farmland preservation, urban gardening, and the siting of housing and neighborhood development to minimize transportation and maximize food procurement at the neighborhood level (Lapping and Pfeffer, Chapter 7; Freedgood, Chapter 6). Tax and lending policies support this.

- *Urban, suburban, and rural areas are all food secure.* They contain farmers' markets, private and community gardens, and supermarkets. Low-income families do not depend on charity from food pantries and soup kitchens. A notion of *comida* exists—the Latin American cultural precept of community created by eating in a particular social context (Esteva 1994). There are neighborhood food sites (possibly schools) where families of all income levels rotate cooking for each other on a weekly or monthly basis.

- *Global trade occurs in a context of conserving energy and optimizing self-*

reliance for every country, that is, to the level that any country can produce food security for its population. The agricultural sector is protected in all countries to ease hunger and enhance local economies.

- *Citizens are interested in the larger food system and their connections to farmers, ranchers, and other parts of the food chain.* As the distance between consumers and producers is shortened, consumers have more power to know and influence the quality of food (Berry 1993). People care about food and the linkage it makes between them and the natural world. They know how to prepare food and are interested in eating more whole foods. They make consumption choices with much more knowledge about how choosing imitation and synthetic foods affects health and economics; they are willing to take responsibility for their own food choices; they recognize that how and what they choose are social and moral issues (Miller 1995).
- *The economic system acknowledges the cost of producing and distributing food in an environmentally destructive manner and does not discount the future consequences of present consumption.* Food prices internalize these costs but offset them by lower advertising and transportation bills.
- *People are willing to pay reasonable prices for food.* They understand that the price of food should cover the costs of producing it and recognize that their purchases can contribute to the economic viability of local communities. They understand that whether food is "cheap" depends on one's income and that food is not cheap for people receiving excessively low wages and welfare benefits.
- *Economic concentration and advertising are reduced.* Food marketers compete on a more equitable basis by providing all the information desired by food purchasers, and consumer sovereignty is restored.
- *Everyone, from farmers to politicians, is better educated about food, nutrition, and the food system.* Food becomes a useful vehicle for civic literacy. At the secondary and university levels, students studying in all areas of food or agriculture take courses in food systems and food ecology and are part of multidisciplinary research teams. Students are encouraged to undertake experiences in food systems at the farm and community levels.

Attention to Public Policies and Values

As idealized as the above list may seem, almost every piece of it is being addressed by a project somewhere in the country. Farmers' markets and local supermarket/farmer collaborations are most visible; there also are community-supported agriculture operations, various farmland preservation programs, and alternative marketing and value-added projects, including farmer-owned processing plants. Farmer co-operatives in the Northeast are increasing in number (Hilchey 1996). Larger institutions are mounting projects to buy more foods produced locally. The Chefs Collaborative and other activities in the food-service industry are providing new markets for small growers and enhancing consumer knowledge of local food. Small food businesses are starting up in many places, and a few community and economic development corporations see the benefits of paying attention to food enterprises. Researchers are developing machines for small food-processing plants that are of an appropriate scale and conserve resources. Finally, in food-system planning bodies in a dozen places around the country, people from different parts of the system are sharing information and improving connections among themselves, their businesses, and local governments.

These are the "pilot" projects for the new vision, and many people engaged in them across the country have come together to share ideas and learn from each other. Yet they also are fully aware that the vision will be achieved only through many more "processes and relationships that create the conditions of community and ecology" (Kneen 1989, 140). Many of the institutions making up the present food system will also need to change, as will the laws that govern them. Progress requires both private and public efforts, along with policy change at many levels. The system will not become more connected if, for example, agriculture and food policy are not themselves better integrated. Any success in bringing sectors together can be overturned by policies that force them to operate separately. Most important, food consumption needs should heavily influence production decisions. This idea seems to have little currency now, when competing for exports seems to be our main national agricultural policy goal, but at the local level, and in the future when fuels for transportation are limited, its soundness will become more apparent.

Policymaking in *all* arenas needs to be much better integrated, and much greater attention must be paid to issues that cross different sectors of the food system. Antitrust laws, taxes, and farm subsidies should not encourage greater economic concentration but should lower barriers to entry for farmers, processors, and distributors.

Also, food-safety regulation should not foster bigness as it does now. If markets operate more perfectly, consumers will have more information about where and how food is produced. If suppliers are closer to home, shoppers can know more about them and through familiarity will generate pressure to maintain higher safety standards. To support such ideas, laws and regulatory decisions should allow inclusion of social and economic criteria. Regulations should not induce such high expenditures that smaller-scale, local food businesses cannot compete.

It may seem premature to outline the policy changes that would be needed to achieve the vision I have offered because governmental actions likely will occur only when enough citizens support and encourage them. Unfortunately, the U.S. food system suffers from a problem that Kramer (1990, 99) has termed "public unheedfulness." Unlike many Europeans, most Americans have never paid much attention to the implications of their food choices. Most consumers, including farmers in their role as consumers, no longer see food as the source of health and nurture. To build an alliance between farmers and others will require a change in attitude regarding both the place of food in the life of individuals and communities and the role that food can play in generating community.

Yet this is likely to occur only if enough people choose to reformulate their values or reclaim values that have been neglected. To rehumanize consumption, reintegrate food into the culture, and turn producers and eaters into allies, people will have to acknowledge and act on their responsibility to the common good and the need to balance it with self-interest (Browne et al. 1992). They will need to examine carefully the effects of their (mainly) passive acceptance of the present food market and decide to go into or develop markets in a more active, empowered way. To create the conditions of community and ecology, consumers will have to "choose to have less choice" (Miller 1995).

This is not an impossible task if people motivate themselves by recognizing that, among other things, much of the choice in the food market is spurious, an illusion of artificial colors, flavors, and other additives; wasteful, particularly of the fossil fuels used to transport foods all over the world; and harmful to local farmers and communities. On the positive side, choosing more of the family's food supply from closer sources provides a plethora of its own benefits. The pleasure of eating is greatly enhanced by consuming fresh, healthful, flavorful food. There is an added pleasure, as Berry (1990) points out, in knowing that the producers of the food have considered how the natural world functions, have minimized soil and water degradation, and have treated animals humanely. In the new food system, people will continue to consume imported foods, but it will be an educated choice, based on information and reflection. Back home, renewed trust among farmers and eaters will also add to the pleasures of eating while it builds friendship and community and helps local economies. Caring and responsibility will be the hallmarks of farmer and consumer/citizen relationships; these in turn will enhance stewardship and cooperation far into the future.

Reference List

Berry, Wendell. 1990. "The Pleasures of Eating." In *Our Sustainable Table,* edited by Robert Clark, 125–31. San Francisco: North Point Press.

———. 1993. *Sex, Economy, Freedom, and Community.* New York: Pantheon Books.

Browne, William P., Jerry R. Skees, Louis E. Swanson, Paul B. Thompson, and Laurian J. Unnevehr. 1992. *Sacred Cows and Hot Potatoes: Agrarian Myths in Agricultural Policy.* Boulder: Westview Press.

Clark, Robert. 1990. Preface. In *Our Sustainable Table,* edited by Robert Clark, ix–xi. San Francisco: North Point Press.

Connor, John M., Richard T. Rogers, Bruce W. Marion, and Willard F. Mueller. 1985. *The Food Manufacturing Industries: Structure, Strategies, Performance, and Policies.* Lexington, Mass.: Lexington Books.

Cronon, William. 1991. *Nature's Metropolis.* New York: W.W. Norton & Co.

Culinary Institute of America. 1996. "Philosophy of Food and Cooking." In *Greystone, The Culinary Institute of America, Napa Valley.* Education Programs, January to December 1996.

DeLind, Laura. 1992. "Cheap Food: A Case of Mind over Matter." Paper presented at the Annual Meeting of the Agriculture, Food and Human Values Society and the Association for the Study of Food and Society, June 4–7, East Lansing, Mich.

Esteva, Gustavo. 1994. Re-embedding food in agriculture. *Culture and Agriculture* 48:2–13.

Flora, Cornelia. 1992. "Sustainable Agriculture, the Structure of Agriculture, and Rural Communities." In *Alternative Farming Systems and Rural Communities: Exploring the Connections.* Proceedings of the Institute for Alternative Agriculture Ninth Annual Scientific Symposium, 39–48. Greenbelt, Md.: Institute for Alternative Agriculture.

Goodman, David, and Michael Redclift. 1991. *Refashioning Nature: Food, Ecology and Culture.* London and New York: Routledge.

Goodman, David, Bernardo Sorj, and John Wilkinson. 1987. *From Farming to Biotechnology: A Theory of Agro-Industrial Development.* Oxford: Basil Blackwell.

Hilchey, Duncan. 1996. Northeast grower's coops making a comeback. *Farming Alternatives* Winter: 8–9, 14.

Kneen, Brewster. 1989. *From Land to Mouth: Understanding the Food System.* Toronto: NC Press Limited.

Kramer, Mark. 1990. "Are Farmers an Endangered Species?" In *Our Sustainable Table,* edited by Robert Clark, 95–104. San Francisco: North Point Press.

Miller, Daniel. 1995. "Consumption as the Vanguard of History." In *Acknowledging Consumption: A Review of New Studies,* edited by Daniel Miller, 1–57. London and New York: Routledge.

Smith, Stewart. 1992. "Farming Activities and Family Farms." Presentation at the Joint Economic Committee Symposium on Agricultural Industrialization and Family Farms: The Role of Federal Policy, October 21, Washington, D.C.

Van En, Robyn. 1995. Eating for your community. *In Context* 42 (Fall): 29–31.

5

Agricultural Landscapes in Harmony with Nature

Joan Iverson Nassauer

The Popular Image

The popular image of the countryside is a visual metaphor for human life in harmony with nature. The popular image has enormous appeal, but the reality of ecological health in agricultural landscapes sometimes contradicts the image. This image is more likely to be found in children's books than in pesticide commercials; it is more likely to be seen in a picture hanging on your wall than on a drive through central Illinois; it is more likely to be a part of the view from a home newly constructed on converted farmland than from a home constructed ten years ago and sitting in a sea of five-acre lots.

You know this image: A mix of crops weaves a varied field pattern, livestock graze on the land, woodlands and streams make sinuous borders along the fields, tidy farmsteads dot the landscape. There are fish in the pond, birds in the sky, and wildlife in the woods. The air is clean. There is a small town nearby with a school, stores, and churches. You might not live in this landscape, but you would like to visit it, and when you did, you could stop and enjoy a friendly talk with the farmer and buy fresh produce you couldn't buy in the city.

This image could be called generic nostalgia, but that is only evidence of its broad recognition and enormous appeal. For most Americans this image embodies the same values and expectations they will support in a new vision of the American landscape. These values and expectations include the following:

- The countryside is inhabited by friendly people who enjoy farming and are good stewards of the land.
- People are safe and welcome there.
- The countryside is clean, unpolluted, and uncrowded.

59

- It produces healthful food—better than what you can buy in the supermarket.
- The countryside provides habitat for wildlife in a way that is more natural than the city.
- The countryside is an attractive place to visit. You can drive through and enjoy the scenery.

In summary, the popular image of the American agricultural landscape grows from beliefs that the countryside is ecologically and socially healthy. It also reflects a belief that even those who do not own farmland or live on farms belong in the countryside as welcome visitors to appealing landscapes. It may not be unreasonable to expect these beliefs to be matched by reality. In fact, the popular image creates a demand for this reality.

A New Vision: Ecological Health in Beautiful Nature

The new vision of American agriculture should grow from the core of this popular old image. As we learn more about the complex ecologies of all landscapes, harmony with nature becomes more of an imperative than an ideal. The common belief that the countryside is a form of nature coincides with the growing awareness that we must ensure ecological health in the countryside. This makes it possible for popular expectations to propel public policy into achieving ecological health for American agricultural landscapes.

The image of nature in the countryside is at the heart of the beauty people seek and find there. When we construct a new vision of agricultural landscapes, we would be foolish to ignore the cultural power of this image. America has become a suburban nation, and some rural counties have grown wealthy as urban people have ventured from cities in search of beautiful nature in the countryside. Beauty could be the sole focus of a new vision. Countryside landscapes are integral to the quality of life in metropolitan areas and are the basis for thriving rural economies. But focusing on beauty alone tends to leave both beauty and the larger agricultural and ecological functions of the landscape undefended. In our culture, aesthetics is mistakenly denigrated as superficial while advertisers construct beguiling images to manipulate our behavior. A thoughtful strategy for American agriculture will use the power of images—not to manipulate but to communicate the ecological achievements of agricultural policy. The agricultural landscape doesn't need a billboard of nature to make us feel that the water is clean and the food is good. The agricultural landscape advertises itself when it conveys the popular image of nature.

Those who track the loss of farmland to exurban development, or fear the pollution of water by feedlots and fertilizers, or know the surreal sterility of monoculture in the grain belt see the jarring contrast between the image and the reality. But it is just this contrast that sets up the possibility for a new vision. If we didn't expect the countryside to be natural, if we saw it as only another form of industry, we might be complacent about the inevitability of lost habitat. If we did not personally enjoy the appearance of good stewardship on the land, we might see the deteriorating health of the countryside as someone else's problem. But because we expect agriculture to be in harmony with nature, we hold a collective image of what the countryside is and should be. A new vision can bring the ecological reality of American agricultural landscapes closer to the evocative popular image of a beautiful countryside.

Limited Knowledge and Intelligent Tinkering: Ecological Conservatism

The popular image of harmony with nature is not enough to tell us what will actually work to improve the ecological effects of agriculture in the future. The image is a democratic goad to healthy agricultural landscapes, but it is not an instruction kit.

We know that new agricultural landscapes cannot be simple replicas of the past and that the agriculture of the future cannot be based on the limited insights of a single discipline or a narrowly construed scientific method (Gerber, Chapter 12; Lacy, Chapter 15). At its best, scientific understanding gives us a conceptual overview of flows of water, soil, nutrients, chemicals, and plant and animal species through the agricultural landscape and its products. It also partly explains the movements of people in and out of the countryside and their reasons and means for staying there. But science does not give us a definitive understanding of how particular agricultural landscapes work, and it cannot give us the kind of informed permission for wholesale disturbance of functioning ecosystems and communities that some developers and agriculturalists desire. Rather, many scientists have reached a conclusion that parallels that of concerned skeptics, a message of caution: Be careful when you change a landscape that works; be cautious about unintended effects of your technology; conserve what works when you experiment with what might work better. Aldo Leopold's (1966) dictum still holds: The first rule of intelligent tinkering is not to throw away any of the parts.

Caution does not prevent change. Rather, it may lead us to reflect on the scale of change that has been introduced into North American landscapes over the past century and to amend that sweeping scale in agricultural landscapes in the next century. Caution may lead us to return some elements that bring variety and ecological balance to the homogeneous patterns of production in agriculture. As Julia Freedgood describes in Chapter 6, the agricultural landscape *must* function ecologically. Agriculture is integral to large-scale processes of energy conversion, aquifer recharge, soil development, and water and habitat quality. Undoubtedly, we must move to restore some beneficial effects of old agricultural practices that left large-scale ecological processes intact. This does not mean mindlessly returning to old ways. It means critically selecting what worked in the past and inventing new patterns that will work now and in the future. It suggests we begin tinkering not by reconstructing the landscape as our grandparents farmed it or as Europeans first encountered it, but by reexamining some old farming practices and indigenous ecosystems. This perspective on agricultural research resembles the assumptions underlying experiments with cultivating perennial grains more than the assumptions underlying pesticide development.

Knowledge of the ecology and culture of a countryside that works may inform us, but that knowledge is incomplete. It cannot tell us what to do. Caution may lead us to a form of conservation that acknowledges our collective hubris. More accurately than the dichotomous stewardship terms of *conservation* and *preservation, ecological conservatism* describes a way of farming that is attentive to what we do not know. With ecological conservatism, past landscape patterns that were successful in maintaining large-scale ecological processes would be the primary guide for inventing new landscape patterns. Consistent with the conclusion that we cannot predict all the effects of landscape change, ecological conservatism would suggest that variations on the patterns that work be introduced gradually—at small spatial and temporal scales—and monitored for their effects. Such small-scale monitored experiments suggest an approach like the on-farm research that has become an integral part of sustainable agriculture. Of course, "small" and "gradual" are relative terms. But if we use energy consumption as one measure, we can be certain that industrial agriculture, with its attendant use of fossil fuels, has brought us too far too fast.

Intelligent tinkering gives us a mechanical metaphor for ecological quality. We imagine the tinkerer at the workbench, with all the parts laid out to experiment with how they might fit together to serve a purpose. The machine the tinkerer makes is not necessarily elegant, but it works, and the extra parts have been carefully saved for the next time they might be needed. Someone entering the workshop might not see the order in the parts lying here and there, but the tinkerer knows where everything is.

To use the popular image of the countryside to advance ecological health, intelligent tinkering must be translated into visual terms. The new vision of the agricultural landscape must portray ecological health by drawing on what people already recognize. Our cultural image tells us what nature in the countryside looks like. We can adapt this familiar visual metaphor to portray ecological health (Nassauer 1992).

The Look of the Land: Knowledge and Image

Knowledge and image must be intentionally meshed by those who care about public support for the ecological health of the agricultural land-scape. As the agricultural landscape recovers characteristics that support ecological health, who will notice, and who will know? Who will advocate, and who will pay? Because image is a reflection of cultural traditions rather than critical analysis, many people will not perceive ecological gains unless the agricultural landscape looks healthy. Because many characteristics that support ecological health are invisible or difficult to see or even contradict the image, knowledge and image will not inevitably converge in the look of the land. They must be designed so that image matches knowledge.

When we don't see ecological health, it isn't because we aren't looking. In Western culture, at least since the seventeenth century, people have entertained themselves by looking at the land to judge the wealth and character of the landowner and to enjoy the beauty of the scenery. In the eighteenth century, educated Europeans and Americans began enjoying the look of the landscape for what it told about the natural history of a place. Although that pastime spawned the natural sciences and the conservation and preservation movements, it has remained a rarefied pursuit. Most people say they enjoy nature, but few people can identify plants or animals. Nonetheless, driving for pleasure is the most popular form of recreation among Americans. When we take a drive in the country, we may not know what we see, but we expect to like it.

Among farmers and homeowners the idea that the way your place looks reflects on you is such a commonplace that people seldom talk about it, but we all know it and think about it as we drive through our neighborhoods. Aldo Leopold (1939) went so far as to state that "every farmer's land is a portrait of himself." The fact that views of agricultural landscapes are redolent with messages about their caretakers serves to remind us that nature in the countryside is always about people and how they take care of their places. The look of nature so thoroughly infused with human intention should not be confused with nature in the wilderness. Each evokes a different image. Nature in the wilderness may be sublime. In the countryside, the sky and weather bring events of sublime grandeur, but the land is tended.

For knowledgeable viewers, the landscape tells a story that is animated not only by people but by processes and events. Farmers know the drainage and soils and slopes of their land with a subtlety that often surpasses science or engineering. Hart (1975) coined the term "look of the land" to suggest all that the appearance of the landscape can tell us about its history and use. Watts (1957) invited us to "read the landscape" for clues to its ecological character. Lynch's (1960) primer for designers and planners instructed that the environment is an enormous communications device, and he demonstrated how the landscape could be designed to evoke a sense of place. New agricultural landscapes should communicate information so that people can become more knowledgeable about the history and ecological function of the landscape. The landscape should communicate in the most recognizable terms, melding the popular image with cues to ecological function.

Careful Change: Melding the Popular Image and Ecological Knowledge

Melding image and knowledge in public perceptions of the agricultural landscape will lead us to ask two questions whenever we introduce change to increase ecological health. The first, most fundamental, question is: What change will increase ecological health? In Chapter 6, Julia Freedgood describes several potential changes, from increasing connectivity of uncultivated patches to reducing the use of herbicides. Landscape ecology (Forman 1995) and conservation biology (Meffe and Carroll 1994; Noss and Cooperrider 1994) suggest a rapidly expanding set of principles for change. The second, more strategic, question is: What do people expect that kind of ecological health to look like? This is different from asking what that kind of ecological health actually looks like. It is a question of how we portray nature to fit into a cultural tradition. It is a question of how we take care of our landscapes so that our neighbors will admire and enjoy the nature there. To answer, we begin with how people commonly look at the landscape, not how they might be educated to see it.

For most people the expected image of nature in the countryside more closely matches the appearance of a garden or a park than the wilderness. Nature in the countryside means fields and woodlands, birds and flowers, streams and ponds, hills and valleys, barns and live-stock, and very few houses. It is nature at a comprehensible scale, where fences and hedgerows run between fields, trees grow beside streams, and a person could walk from here to there. This is an inhab-ited nature that invites human involvement, kept neat by those who live there and watch over it. It is nature enhanced by signs of human tend-ing, from freshly painted fences and buildings to straight rows in weed-free fields (Nassauer 1988). All this creates the possibility and desir-ability of a type of nature that looks quite different from what we expect to find in the wilderness.

Knowing that this is the way people expect nature in the countryside to look should affect our practices and plans for the new agricultural landscape. Those who have worked to restore habitat in agricultural landscapes know that even people who enjoy nature often object to the uneven, weedy appearance of habitat plots, restored wetlands, or re-serve parcels. This is not the tended nature they expect. In the past decade, wetlands and prairies have sometimes been obliterated in part because their ecological quality was not apparent to those who saw them. Nature and ecological health will not be unified in popular per-ceptions unless we plan it that way.

Intelligent tinkering at small scales in agricultural landscapes might monitor many ecological effects, asking, How does this new practice af-fect water quality, species diversity, or economic productivity? It should also monitor perceptual effects, asking, Does the look of the landscape communicate its ecological quality? Do people enjoy the nature they see here?

Familiar Patterns and Intelligent Tinkering: Recognizable Beauty

Making new agricultural landscapes recognizably beautiful will require working with the familiar landscape language of the popular image. Fa-miliar language does not prevent new statements. It allows parts of the language to be used in new ways. Old, recognizable patterns can be used to signify what new ecological elements mean. Below, beginning at the broadest scale, are some possible ways that ecological changes could be juxtaposed with familiar characteristics of attractive land-scapes. Large-scale patterns, edges between types of land cover, and frames around ecosystems that introduce new biodiversity—all are strategies for using what is inherently attractive to convey what actually is healthy in the landscape.

Pattern: Panoramas and Patches

People value panoramic views of the countryside. Panoramic views are more beautiful if they include distinct patterns created by the edges of fields and forest, as with riparian reforestation. In this example (Figure 5.1), the existing landscape has a narrow wooded strip along a river running through the wheat fields of North Dakota. In the simulated alternative (Figure 5.2), ecological function is enhanced by larger wooded patches that transect lowland to upland ecosystems and establish continuous cover along the riparian corridor. Having many fields in a single landscape, such as those created by strip cropping and crop rotations, has the same effect: creating distinct edges and an overall pattern that is associated with good stewardship. Where panoramas are more extensive and the patterns more varied because they look out over rolling hills, they are even more beautiful (Figure 5.3). Where elevated viewpoints create opportunities for panoramic views over rolling hills, new landscape patterns that create more small fields, more crop variety, more hedgerows, and more wooded patches will help the public see nature in the countryside.

Figure 5.1. Beyond a narrow, intermittent fringe of riparian vegetation, the landscape within the river corridor displays little variety in land cover. (USDA-NRCS)

Figure 5.2. This simulation of the same river corridor shows how riparian reforestation that creates some relatively large woodland patches and connected cover along the river also creates a vivid pattern of varied land cover. (Video imaging: Robert Corry)

Figure 5.3. Stripcropping and crop rotations also create distinct visible patterns that are associated with good stewardship. The rolling terrain where stripcropping is practiced creates panoramas that make the patterns even more apparent. (USDA-NRCS)

Edges: Curves and Buffers

People expect nature to have curved edges. Several traditional soil and water conservation practices, such as contour plowing, strip cropping, and terracing, emphasize the curve of the land and vividly convey good stewardship, as shown in the video imaging simulation in Figure 5.4. New conservation and reserve plans should use vivid patterns and curves where they fit the land. Conservation reserve parcels that incorporate no pattern are difficult for viewers to decipher; they have been mistaken for weedy fields (Figure 5.5). The simulation in Figure 5.6 shows how maintaining visible curved edges and a pattern of different landcover types can be part of new conservation plans that might include riparian restoration and reserve parcels. Sustainable agriculture practices should be applied with the same awareness of visible pattern and edges. For example, the pattern of paddocks in a landscape managed for rotational grazing can have the same effect as strips.

Where streams, ponds, and roads can be allowed to meander or follow the curves of hillsides rather than slicing through the landscape on a straight line, people will see nature. Where habitat enhancement or the need to buffer the flow of nutrients or sediment suggests planting strips of perennial cover through fields and along streams, the strips can be broadened and curved to convey the image of nature.

Figure 5.4. Conservation practices such as stripcropping create edges that curve with landforms. These vivid patterns are associated with nature.

Figure 5.5. Some reserve and conservation practices, such as the conservation reserve simulated in this picture, have ecological benefits but little perceivable pattern. Consequently, they tend to look weedy or neglected. (Video imaging: Regina Bonsignore)

Figure 5.6. The combination of conservation reserve, riparian restoration, and traditional conservation practices in this simulation create strong patterns with curved edges. Several cues to care, such as a mown strip, flowering plants in the reserve seed mix, and a white fence, are also shown in this alternative. Ecological purposes are achieved, and the image of harmony with nature is created.

Frames: Property and Pride

Because we expect countryside nature to be inhabited by good stewards, landscape change should also display human presence as cues to care (Nassauer 1995). Cues to care are reassuring signs of the good intentions and hard work of the people who own a place; the cues make fields and farmsteads look neat and tidy. They include mown strips, flowering plants, and freshly painted signs and fences. For example, in Figure 5.7, the grassed waterway would create a pattern and has a curved edge. In the simulation in Figure 5.8, the same waterway is shown planted with a seed mix heavy with native flowering plants. Habitat value is enhanced and the ecological intent of the practice is even more vividly conveyed. In the Midwest, the color white is associated with a neat, tidy farmstead. In Figures 5.9 and 5.10, the simulations of a stream and its restoration include white fence posts and bird houses to show that someone is taking care of the restoration—it is not abandoned land. Cues to care allow farmers to display pride of ownership. When they are used to frame ecosystems that we might perceive as messy or weedy, the cues help us see the landscape as tended nature, not neglected land.

Figure 5.7. This grassed waterway contributes to the image of harmony with nature because of its bold pattern and curved configuration. (USDA-NRCS)

Figure 5.8. If the waterway were planted with a seed mix of native plants with an abundance of flowering plants, it would contribute further to ecological health and would even more powerfully convey the image of nature. (Video imaging: Regina Bonsignore)

Making Change Popular: A New Agricultural Landscape

In the new agricultural landscape, ecological health and agricultural production will be communicated by a landscape that resembles the popular image of beautiful nature in the countryside. At the same time, the beauty of the countryside will be protected and perpetuated by the ecological and economic functions the countryside performs. In this vision, beauty is more than skin deep, and beauty is not a trivial byproduct of serious policy. Rather, it is an intentional way to achieve popular support for serious ends: ecological health, agricultural production, and quality of life.

The new agricultural landscape will be beautiful in a way that invites tourism. Scenic roads and byways and places for visitors to stay will become more appealing as parts of the countryside that have lost habitat, streams, or a varied landscape pattern regain a more recognizable image of nature. The countryside will be beautiful in a way that promotes the value of agricultural open space as part of the metropolitan fabric and protects urban agriculture.

Figure 5.9. This stream restoration site could be restored in a way that would prevent erosion, filter nutrients, and create connected habitat.

Figure 5.10. By incorporating cues to care, such as mown strips, white fence posts, and birdhouses, the stream restoration will not be mistaken for an abandoned field that needs to be "cleaned up." (Video imaging: Regina Bonsignore)

The new agricultural landscape also will communicate the ecological benefits of the countryside to the body politic. The pleasure of nature in the countryside will portray the good stewardship of farmers for all to see.

Reference List

Forman, Richard T. T. 1995. *Land Mosaics.* New York: Cambridge University Press.

Hart, J. F. 1975. *The Look of the Land.* Englewood Cliffs, N.J.: Prentice Hall.

Leopold, Aldo. 1939. The farmer as conservationist. *American Forests* 45 (6): 296–97.

———. 1966. *A Sand County Almanac.* New York: Oxford University Press.

Lynch, Kevin. 1960. *The Image of the City.* Cambridge, Mass.: MIT Press.

Meffe, G. K., and C. R. Carroll. 1994. *Principles of Conservation Biology.* Sunderland, Mass.: Sinauer Associates, Inc.

Nassauer, J. I. 1988. The aesthetics of horticulture: Neatness as a form of care. *HortSci* 23:6.

———. 1992. The appearance of ecological systems as a matter of policy. *Landscape Ecology.* 6:4, 239–50.

———. 1995. Messy ecosystems, orderly frames. *Landscape Journal* 14:161–70.

Noss, R., and A. Cooperrider. 1994. *Saving Nature's Legacy.* Washington, D.C.: Island Press.

Watts, M. T. 1957. *Reading the Landscape of America.* New York: Macmillan Publishing Co.

Part II

6

Farming to Improve Environmental Quality

Julia Freedgood

Agriculture Today

Agriculture is a fundamental human activity that has evolved along with us, nourished us physically and culturally, and provided a living laboratory for us to develop our knowledge of science, economics, and technology. According to Rick Potts, director of the Smithsonian Institution's Human Origins Program, "cultivation of living things has provoked the growth of human dominion" (Potts 1996, 23). Today, the extent of our dominion has great environmental effects. How we practice agriculture in the future will determine the quality of life for succeeding generations.

With our current level of understanding about natural systems, we can improve farming practices and emphasize those that have long-term benefit for people and the planet. If we declare this to be a public priority, invest more research and extension money in integrated systems approaches, and support the outcome of this research with public policies, agriculture can be a model of how an industry can adapt to environmental demands, support the restoration of natural systems, and encourage long-term economic stability.

Humanity and agriculture as a crucial human activity are part of the environment. As much as industrialization has given us the tools to place ourselves above nature, in trying to control it we have discovered how connected to the environment we really are. Thus, even in a century when people have proved to have a substantial capacity for mass destruction, in a time likely to be remembered for overproduction, consumption, and waste, we can still create beauty, harmony, and balance. And while we will surely continue to reconstruct the world to suit our species, this need not occur at the expense of environmental quality. We can adapt our farming systems to copy nature, not just to con-

77

sume it, and in the process improve the environment in which we live.

Serious resource challenges face the world as we approach the end of the twentieth century: declining per capita food production, climate change, and escalating species extinction. While experts argue over the severity of these threats, several trends are clear. After several million years of evolution, the world population reached 1 billion people at the end of the eighteenth century. Since then our population has grown dramatically to around 5.4 billion and is expected to increase annually by another 90 million, or a billion people every eleven years. Meanwhile, world production of food per capita is thought to be declining. In 1990, the U.S. Department of Agriculture reported that grain consumption had exceeded production for three straight years (Smith 1990). Global emissions of carbon dioxide have been increasing so rapidly that some scientists predict that over the next one hundred years, the accumulation of carbon dioxide and other gases in the atmosphere will increase global surface temperature by several degrees Fahrenheit, disrupting natural and agricultural systems. Finally, worldwide depletion of natural resources is resulting in unprecedented loss of biological diversity. Wilson (1993) estimates that the rate of extinction is now one thousand to ten thousand times higher than in prehuman times. He predicts that if destruction of habitat continues at its present pace, 20 percent of the earth's species will be doomed to premature extinction in the next thirty years.

The United States is endowed with some of the earth's finest natural and agricultural resources. Yet every year we convert about 1 million acres of farmland to urban use and rural development. According to the National Resources Inventory (U.S. Department of Agriculture 1995), between 1982 and 1992, 4 million acres of this was prime farmland. Developed land increased by 14 million acres, with two-thirds coming from the conversion of crop and forest land. Meanwhile, the U.S. population as a whole grew by less than 1 percent annually (Spitze 1990, 13).

Population growth is not the primary cause of farmland conversion. We are consuming vital natural resources to move people from central cities and towns to sprawling suburbs and isolated farming and ranching areas. Stripping the topsoil from farmland and replacing it with pavement for subdivisions and shopping malls transforms a renewable resource into a nonrenewable one. Unmanaged growth not only claims valuable agricultural resources, it destroys wildlife habitat, reduces species diversity, and degrades our air and water.

Although these projections are grim, when I look at the agricultural horizon, I am encouraged by what I see. We have difficult choices to make, hard allegiances to forge, and many obstacles to overcome. But it often takes a crisis to bring people together, and that process has begun.

In the past few years, I have participated in many community consensus-building processes, listening sessions, conferences, and working groups of farmers, environmentalists, policymakers, land-grant researchers, Extension agents, representatives from farm and community organizations, and other stakeholders in the food and farming system. People with diverse interests and frequently conflicting points of view are finding common ground and are beginning to work together. We are finding ways to reconcile—and even combine—opposing ideologies to further a shared purpose, which is a prosperous future for American agriculture.

The belief that agriculture can and should provide environmental amenities is growing. Farmers and ranchers, who have long considered themselves stewards of the land, are increasingly willing to modify environmentally damaging production practices. Environmentalists, who once were hostile toward agriculture, are beginning to declare farming the preferred land use in urban watersheds, as long as whole-farm management or Best Management Practices are applied. State and federal agencies are finding that public acquisition and regulation to protect natural resources cannot succeed without collaboration with private landowners. Community leaders are learning that farmland contributes more to their tax base than subdivisions and are encouraging economic development strategies that support farming.

Agriculture is taking advantage of modern scientific advances in both agronomy and ecology, improving production practices to benefit the environment while continuing to provide a steady supply of food. It can build upon these approaches to do much more. This will depend on increasing public recognition of the importance of agriculture and the many public benefits that it provides. Broad public support will help refocus agricultural policy, research, and education to achieve multiple goals. This is a big challenge, but it is possible. It is up to the American people—urban and rural, farmers and environmentalists, economists and agronomists—to support ecological farming systems.

Since World War II, public policy, research, and education have promoted an industrial model for agriculture. The modern mantra has been to maximize production: get big or get out. This policy has achieved its goal. For fifty years, American agriculture has been applauded for providing a cheap and abundant food supply. But success has come with a price that includes water pollution; soil erosion; diminished genetic diversity; dramatic decreases in farmers, farming, and farmland; and eroded public trust. We need new policies to guide our agriculture in the next millennium.

On the scale of human evolution, agriculture is still a young experiment. I have pointed out the shortcomings of modern policies and practices, but all was not well before industrialization either. Humans have always played a role in species extinctions, and for centuries we have replanted the globe. Agriculture is only one of many industries and activities that have degraded the environment. Despite the enduring Western world view that separates people from nature, modern science has brought us to the point of recognizing that ecology is destiny. We are as affected by the environment as we are effective at disturbing it. The issue is not who is to blame but whether we are learning from the past. I believe we are. I believe we can create new systems to respond to the environmental challenges of the twenty-first century.

An encouraging example occurred in November 1995, when almost two hundred people came together to discuss ways to improve the environment through agriculture. The "Environmental Enhancement through Agriculture" conference, held in Boston, Massachusetts, was organized by the Tufts University School of Nutrition Science and Policy, American Farmland Trust, and the Henry A. Wallace Institute for Alternative Agriculture (Lockeretz 1996). It was premised on the belief that agriculture can do more than mitigate its own negative effects: it can help solve environmental problems. Many presentations focused on alleviating the harmful effects of a production-maximizing agricultural system, but many others addressed ways that agriculture can greatly improve the environment we already have.

Across the country, farmers and their communities are working together to integrate agricultural production with water-quality enhancement and watershed protection (Hall 1996). Farmers are profiting by composting food and farm wastes and converting municipal refuse to soil amendments (Halstead et al. 1996). Farmers, ranchers, and researchers are finding creative ways to provide wildlife habitat and encourage genetic diversity in plants and animals while maintaining agricultural production (Payne et al. 1996; Anderson et al. 1996). Research is underway on perennial grasses and short-rotation woody crops as potential alternative energy sources (Tolbert and Schiller 1996). Municipalities are even beginning to investigate pollution-hungry crops to plant on landfills. Finally, agriculture can maintain attractive landscapes and offer beautiful views, a benefit that is immediately apparent to all.

If public policy, research, and education can succeed so well in developing an agricultural system that *maximizes* production to the point of creating surpluses, they also can be changed to develop farming systems that *optimize* production while supporting both the social and natural environments in which they operate. A synonym for "improve" is cultivate—and cultivation is what agriculture is all about. Thus, in this chapter I present a vision of an agriculture that improves rather than exploits the ecosystems that sustain us. It is premised on the belief that it is possible to combine social and economic goals with conservation so that we can develop agricultural systems that reward farmers, benefit communities, and contribute significantly to resolving the major challenges of our times.

A Vision of the Landscape of the Future

The agriculture of the future will embrace a land ethic that values humanity as a part of the environmental community. Instead of stressing short-term profitability and exploitation, farming will favor principles of equilibrium, with its combined meanings of stability, equality, and balance. And while it still will be highly productive, this agriculture will be concerned more with the efficiency of the whole system than with maximizing yields.

Natural ecosystems are proving to be excellent models for resource management, and agriculture is a good economic activity to teach us how to enhance our environment. An ecosystem approach will minimize waste in the generation of renewable products on the farm, within the landscape, and as part of the larger community. The public will appreciate the fact that culture and commerce occur within these ecosystems and that no matter how urban our lifestyle, we depend on the natural world for life and sustenance. Building on the knowledge we have today, we can farm in ways that more closely mimic nature, applying ecological principles so that farming can improve environmental quality, support biodiversity, and provide sufficient food and fiber to satisfy global needs.

We have learned a lot from the unprecedented scientific and technological advances of our time, and continue to do so. Yet we also will value the artistic and spiritual dimensions of agriculture. We will appreciate nature as much for what it is as for what it can provide. We will develop ways to account for the "soft" and intangible values that elude economists, and mentioning them won't make us squirm.

What will this agriculture look like? First, it will vary geographically: Each locality will have a sense of place, and each region and microclimate will adapt its agriculture to its own environment. Unlike suburban sprawl and shopping malls, the agriculture of the future will feature unique traits in recognizable communities, relieving uniformity with diversity. And while the landscape will appear varied and interesting, standards of use and agricultural practices will share common elements.

Farms will form a mosaic pattern across the countryside, integrating human communities with natural features, such as rivers and streams, forests, and restored wildlands. Whenever possible, farmland will be connected in large blocks instead of being isolated in enclaves surrounded by development. Farms will form a critical piece of a biological and cultural landscape, providing crop and livestock diversity, ecological stability, and community for farmers and neighbors. In this way, they will be part of interacting natural systems, integrated into human development patterns, supporting both the economic and natural environments in which they operate.

Instead of regarding farms as separate economic units within specific towns or counties, we will develop agricultural economies for watersheds and bioregions. Farming will be practiced to improve water quality within watersheds, which supply large populations across local political boundaries. People served by the watershed will invest in supporting agricultural systems that keep their water clean, because doing so is cheaper than municipal water treatment alternatives.

This is already happening. For example, the New York City water-supply system is one of the largest and most complex surface storage systems in the world. In the past, the water received awards for its high quality. But the natural beauty of the watersheds and their proximity to the city made them highly susceptible to land development, which has resulted in high-density subdivisions and increased water pollution. In 1989, the Environmental Protection Agency directed the city to begin filtering its drinking water. A new filtration plant was estimated to cost as much as $8 billion, plus hundreds of millions of dollars a year to operate. But the 1989 Surface Water Treatment rule also set criteria to avoid filtration, so city officials sought an alternative approach that included regulation and land acquisition.

The city initially proposed to eliminate farm runoff by imposing strict regulations on livestock producers. Farmers were outraged that people so removed from them could tell them what to do with their own property. They fought the proposal, arguing that water from farming areas met the EPA's criteria for avoiding filtration, while water from developed areas required treatment.

A task force of farmers and city representatives got together to address the proposed regulations and to foster a creative solution. This group recognized that whole-farm management and implementation of Best Management Practices could be an effective strategy for protecting water quality, and acknowledged agriculture as the preferred land use in the watershed. The parties are now working together to assure high-quality water for the city by supporting good agricultural stewardship (Coombe 1996). Possible strategies include tax incentives and using city money to purchase conservation easements on farmland.

In the future, bioregions and watersheds may gain taxing and regulatory authority to increase their role in determining patterns of growth and development. They could offer financial incentives to support farming practices that enhance water quality, buy conservation easements on critical parcels of land, plant and maintain landscape buffers to separate commercial farming operations from waterways, and encourage ecological planning that would surmount traditional political boundaries. Blocks of farmland will be connected to resource conservation areas and other compatible land uses by greenbelts and traversable waterways. Municipalities will build bike paths and trails to encourage alternatives to automobile transportation. Trees and hedgerows will prevent soil erosion. Drainage ditches will be used as a resource management tool and could become an organizing factor in the landscape. Using a combination of incentives and controls, communities and farmers will work together to eliminate erosion, improve water quality, protect wetlands, and provide wildlife habitat.

Most American food is produced in urban-influenced counties, and this trend is continuing (American Farmland Trust 1994). In the future, municipal and agricultural land uses will support each other better so that metropolitan agriculture can prosper along with its rural cousin. The nature and scale of agriculture will be determined by local market conditions and community demands instead of by federal crop subsidies and federal and state infrastructure policies that encourage conversion of farmland to residential use. Land-use policies will connect the public to the countryside by supporting corridors, greenways, bike paths, rails-to-trails, and other systems. They also will protect farmers from public liability and conflicting development and will maintain buffers and rights-of-way. Farmers will play a greater role in community resource management. New agricultural infrastructure will support integrated farming enterprises that include crop and livestock production, value-added processing, and innovative marketing.

Much is said today about sustainable agriculture and sustainable development and about the need for a restorative economy. But we can go beyond sustaining and restoring—we should be actively engaged in improving. As agriculture has been for most of human history, again it will be the basis of community economics, with management practices designed to maximize long-term value instead of short-term gain. Individual farmers and coordinated community action will complete this vision, and specific government actions can support the transition.

What Farmers and Communities Can Do

Environmentally degrading production practices clearly jeopardize the long-term economic stability of agriculture and farming communities. But it is also clear that Best Management Practices, Integrated Pest Management, and Whole Farm Management systems can be profitable for farmers while providing benefits to the community. In the future, farms will improve water quality, provide wildlife habitat, and convert waste into marketable products, turning environmental problems into profits for the farmer and society. Thus, private landowners will play a larger role in community resource management than they do today, and they will be rewarded for their efforts.

Farmers, other citizens, local boards, community groups, and public officials all will play important roles. Together, they will craft local land-use policies to encourage connections between important farmland and other natural landscape features. These will be treated as preferred uses in natural resource areas, although other land uses may be accommodated that do not conflict with them. Both public and private incentives will support farming practices and land uses that encourage biodiversity and shield fragile natural resources from human development. Trees, wildflowers, and perennial grass buffers will be planted between farms and residential, commercial, and industrial developments to reduce land-use conflicts. The resulting wildlife refuges and bird sanctuaries will complement rather than compete with crops and livestock.

Bioregional growth management will be in force. Equipped with mapping and monitoring technologies that combine socioeconomic and population statistics with natural resource data, state and local governments will team up in strategies for eco-development (both economic and ecological). Agricultural needs will be incorporated into growth-management strategies, and planners and policymakers will be trained to integrate farming into other economic development activities. This will foster beneficial natural-resource employment opportunities for farmers and other resource professionals.

Specific activities will be combined in different places in different ways, but public policies always will reward mutually beneficial approaches. The next generation of resource mapping tools, such as Land Evaluation and Site Analysis and Geographic Information Systems, will be used to inform public policy decisions, incentives, and controls. They will be affordable and widely available to help develop effective whole-resource management practices that are compatible with specific bioregions.

For example, a municipality might encourage combined farm and forest management in a watershed to reduce flooding and sediment damage. Another might develop a cost-sharing program to protect fisheries near farming and forestry enterprises through structural and natural measures, riparian buffer strips, grading, water-spreading diversions, and vegetated waterways. When water is scarce for agriculture, local policies would discourage lawn watering and restrict the application of fertilizers and pesticides on lawns to further protect water quality. Meanwhile, policies will encourage commercial and residential landscaping to be more diversified with trees and wildflowers.

Instead of mowing roadsides, municipalities will plant them in flowers, trees, and perennial grasses, connecting them to the mosaic of other community land uses and complementing other environmental efforts. Planting eco-corridors over gas transmission pipelines, power line rights-of-way, or railroad passageways will make agricultural fields part of an integrated environmental context that many sectors of society will contribute to and benefit from. Municipalities will contract with farmers and other professionals to manage these landscapes. This will create new opportunities for agricultural income, such as allowing farmers to manage grassy corridors with grazing livestock. When communities connect to each other in a bioregional framework, they will create a national system of linked greenbelts, eco-corridors, waterways, and buffers. These would join urban, suburban, agricultural, and wildland areas for low-impact tourism, encouraging alternatives to automobile travel and creating a host of opportunities for people to fall in love with their landscape. A collaboration of constituencies will find an acceptable and dynamic balance among farming and forestry, national parks and wildlands, rural economic development, and metropolitan growth.

Farmers will profit by taking advantage of these closer relationships. Income from production agriculture will be enhanced by other economic opportunities, such as nature-based tourism and farm vacations. Using current marketing approaches, farmers could develop U-pick operations to face greenways and corridors, selling farm-grown products at restaurants and roadside stands along the way. With increasing public support, farmers will cultivate prosperous relationships with their nonfarm neighbors through farmers' markets, community-supported agriculture, and other direct-marketing alternatives. And as integrated-resource managers, they could market a variety of environmental and recreational amenities that they may already be providing free to friends and neighbors, as long as their commercial farming operations are sufficiently protected so that these activities would be complementary and not conflicting. Activities could include providing resource-management services to support hiking, skiing, bike paths, and horseback riding or charging access fees for camping, hunting, fishing, or bird-watching on their land.

New crops and practices will increase progress that is already being made in establishing environmentally friendly farming systems. Composting will be a lucrative agricultural activity, with enterprises converting community food waste, municipal leaves, manure, and other waste products into high-quality fertilizers and soil amendments. These could be applied on farms, but they also could be packaged and mass-marketed to increase farm income. Raising worms and beneficial insects may prove viable farm enterprises.

Energy crops may also find a niche. Perennial grasses and woody biomass crops provide crop diversity, and with it both economic and environmental benefits. They improve soil quality and stability, reduce chemical runoff, provide cover for wildlife, and require lower inputs of water and energy than many other cash crops. These crops can stabilize erosion-prone soils that are not suitable for cultivation of traditional commodities such as corn and soybeans. Furthermore, the development of alternative crops for fuel will reduce our dependence on foreign oil and decrease production of greenhouse gas emissions (Tolbert and Schiller 1996).

Farmers will be trained to improve water quality and build soil organic matter. As a result, they will be able to earn additional income from managing municipal natural resources. They may provide environmental services to individuals and communities, such as resource analysis or scouting and monitoring services for integrated pest management of municipal and suburban landscapes. Although it is too soon to know which new kinds of ecological enterprises will be profitable, the important point is the development of symbiotic relationships in which agriculture enriches the environment.

Our current emphasis on monoculture will be replaced by crop and livestock systems that reflect a more comprehensive understanding of natural ecosystems. Farmers will use perennials for cover crops and nutrient enhancement and will rotate grains, vegetables, and livestock. They will look for alternative enterprises to add to the mix, such as aquaculture. Rotational grazing systems will predominate for sheep and cattle. And while farmers still will be primarily engaged in managing resources to produce food, fiber, and fuel, they will reap financial rewards for environmental services.

Instead of planting crops on marginal farmlands, farmers will grow perennial or prairie grasses and woody crops to attract wildlife, reduce soil erosion, produce fuel, or act as buffers between communities, wildlands, and agriculture. By cultivating natural relationships among wildlife species, farms will attract beneficial insects and birds. Farmers will improve wildlife habitat and reduce pest levels on agricultural land by appropriately managing wetlands and croplands. Modest adjustments in field practices will attract owls, eagles, and hawks by providing food and cover for voles and rodents, which the raptors will keep under control without sacrificing cash crops (Fitzpatrick 1996). Instead of burning rice lands, farmers will flood them in the winter to control weeds and plant diseases while supporting supplemental habitat for waterfowl and other birds (Payne et al. 1996).

These ideas and images are only a sketch of an integrated, community-based, ecological food and farming system. They reflect what is already possible; they will unfold within a context of both global and local conditions. To begin to achieve them, we must work together to secure a strategic base of high-quality farmland, enrich the soil, and restore air and water quality. We must support profitable agricultural systems that sustain genetic diversity, involve people and communities, encourage a mosaic of complementary land uses, and recycle both municipal and agricultural wastes.

How Governments Can Support the Change

We must replace the "bigger is better" notion with new research, education, and public policies that promote an agriculture that is simultaneously ecological, profitable, and responsive to community demands. We must replace policies that encourage growth for the sake of growth in both municipal development and farm operations. We must stop subsidizing sprawl and start managing growth within broadly defined ecosystems. One solution would be to create special agricultural districts that specifically encompass prime, unique, and statewide important farmland or farms with unique microclimates and special historical, cultural, or environmental value. Tax relief, green payments, and other incentives; streamlined regulations; dedicated funding to purchase conservation easements; and new market opportunities could support farmers within those areas so that the agricultural economy will prosper and flourish.

Municipalities must stop relying on the real property tax as their main revenue source. The current tax system encourages premature conversion of farm and forest land, which places more pressure on the tax base and threatens the quality of public education and other services. Tax policies should shift to a system that externalizes social costs so that most revenue would be obtained by taxing harmful actions, such as consuming nonrenewable resources, generating waste, and degrading the environment or public health. Tax relief, environmental regulation, and incentive programs should encourage public benefits and discourage social costs.

Federal farm policy should continue to incorporate ecological goals, supporting environmentally friendly farming practices and discouraging those that deplete natural resources in search of short-term financial gain. Federal soil and water protection will be combined with other natural resource protection efforts. The Federal Agricultural Improvement and Reform Act of 1996 is a significant step in this direction. This farm bill offers new and improved conservation provisions, including farmland protection, and addresses non–point-source pollution, wildlife habitat, and livestock waste management.

Instead of being viewed as a challenge to the status quo, conservation will be considered conservative. Conservation planning will be streamlined because budget cuts will force federal agencies to improve communications systems and coordinate their activities. For example, the USDA's Natural Resources and Conservation Service might work with the Forest Service to combat the fragmentation of critical natural landscapes.

In the future, federal support for agriculture should promote activities that balance food security strategies with improving environmental quality. Mainstream agricultural research and extension will follow an ecological mandate, providing access and adding value to extensive information systems and offering technical assistance on ecosystem-based management.

Public funds will not be used to support highway projects that break up connected tracts of high-quality farmland, dislocate farm communities, or lead to sprawl. Unnecessary conversion of farmland will be reduced; abandoned prime and unique farmlands will be restored. Governments will encourage production practices that optimize economic and environmental efficiencies and maximize public and private benefits, while enforcing regulations to control environmentally damaging practices. In sum, policies will be designed to reward agriculture for providing long-term social goods and services so that it is more economical to improve the environment than to damage it for short-term gain.

Concluding Thoughts

When he won the Nobel Prize, William Faulkner said in his acceptance speech: "I believe man will not merely endure: he will prevail. He is immortal, not because he among creatures has an inexhaustible voice, but because he has a soul, a spirit capable of compassion and sacrifice and endurance." I believe agriculture, too, will do more than just survive; it will prevail.

Agriculture will thrive in the twenty-first century by combining biological and economic principles, replacing the industrial model of maximizing output with an ecological model that supports nature even as it seeks to make nature more hospitable to a burgeoning humanity. Farms and their natural and human communities will become more interdependent. More people will have access to the countryside, and with that access, they will love farms for the food, landscape amenities, and environmental benefits they provide. They will show their support by backing farm-friendly land-use policies, buying food and other farm products directly from farmers, hiring farmers to provide a range of ecological and land-management services, and visiting farms for educational and recreational opportunities. In turn, farmers will support their communities by composting municipal wastes, improving environmental quality, providing educational opportunities, and supporting the local economy. They will produce food, fuel, and fiber in integrated-resource systems that build soil fertility, improve water quality, provide wildlife habitat, and support biodiversity. Farmers and their neighbors will prosper from the symbiosis.

Ultimately, our need to create is as fundamental as our need to consume. While the twentieth century may be remembered for unrivaled exploitation and consumption of the earth's resources, I believe we are gaining momentum in restoring natural systems and encouraging sustainable development. As we deal with scarcity in the future, we must place more value on good stewardship to create a balance that provides for us in the present and prepares us for the future. People working together can connect our economies and communities to nature. We can maintain the environment while adapting it to support us better. We can present a brighter outlook for our children.

Agriculture can lead the way.

Reference List

American Farmland Trust. 1994. *Farming on the Edge: A New Look at the Importance and Vulnerability of Agriculture Near American Cities.* Washington, D.C.: American Farmland Trust.

Anderson, John H., Jennifer L. Anderson, Richard R. Engel, and Bruce J. Rominger. 1996. "Biodiversity within Intensive Farming Systems of the Sacramento Valley." In *Environmental Enhancement through Agriculture,* edited by William Lockeretz, 95–102. Medford, Mass.: School of Nutrition Science and Policy, Tufts University.

Coombe, Richard I. 1996. "Watershed Protection: A Better Way." In *Environmental Enhancement through Agriculture*, edited by William Lockeretz, 25–34. Medford, Mass.: School of Nutrition Science and Policy, Tufts University.

Fitzpatrick, Kerry J. 1996. "Birds of Prey and Their Use of Agricultural Fields." In *Environmental Enhancement through Agriculture*, edited by William Lockeretz, 103–12. Medford, Mass.: School of Nutrition Science and Policy, Tufts University.

Hall, Dennis W. 1996. "Operation Future: Farmers Protecting Darby Creek and the Bottom Line." In *Environmental Enhancement through Agriculture*, edited by William Lockeretz, 1–8. Medford, Mass.: School of Nutrition Science and Policy, Tufts University.

Halstead, John M., Terri Emmer Cook, and George O. Estes. 1996. "On-Farm Composting of Food and Farm Wastes: Economic and Environmental Considerations." In *Environmental Enhancement through Agriculture*, edited by William Lockeretz, 183–91. Medford, Mass.: School of Nutrition Science and Policy, Tufts University.

Lockeretz, William. 1996. Preface. In *Environmental Enhancement through Agriculture*. Medford, Mass.: School of Nutrition Science and Policy, Tufts University.

Payne, Jack M., Michael A. Bias, and Richard G. Kempka. 1996. "Valley Care: Bringing Conservation and Agriculture Together in California's Central Valley." In *Environmental Enhancement through Agriculture*, edited by William Lockeretz, 79–88. Medford, Mass.: School of Nutrition Science and Policy, Tufts University.

Potts, Rick. 1996. *Humanity's Descent: The Consequences of Ecological Instability*. New York: William Morrow & Co.

Smith, Mark E. 1990. *World Food Security: The Effect of U.S. Farm Policy*. Agricultural Information Bulletin No. 600. Washington, D.C.: Economic Research Service, U.S. Department of Agriculture.

Spitze, Mark E., ed. 1990. *Agricultural and Food Policy Issues and Alternatives for the 1990s*. Urbana: University of Illinois.

Tolbert, Virginia R., and Andrew Schiller. 1996. "Environmental Enhancement Using Short-Rotation Woody Crops and Perennial Grasses as Alternatives to Traditional Agricultural Crops." In *Environmental Enhancement through Agriculture*, edited by William Lockeretz, 209–16. Medford, Mass.: School of Nutrition Science and Policy, Tufts University.

U.S. Department of Agriculture. 1995. *National Resources Inventory: A Summary of Natural Resource Trends in the U.S. between 1982 and 1992*. Washington, D.C.: Natural Resources Conservation Service, U.S. Department of Agriculture.

Wilson, Edward O. 1993. "Biophilia and the Conservation Ethic." In *The Biophilia Hypothesis*, edited by Stephen R. Kellert and Edward O. Wilson. Washington, D.C.: Island Press.

7

City and Country: Forging New Connections through Agriculture

Mark B. Lapping and Max J. Pfeffer

The country versus the city. This is a theme older than the nation itself. The conflict remains a potent one, deeply rooted in nearly every aspect of American cultural life. Such distinctions are, of course, stereotypes built upon gross generalizations and oversimplifications that tend to mask the real divisions and cleavages that exist between people, classes, and interests in any given place. Born in the country, the United States became urban with the turn of the twentieth century. Now a suburban nation, the American community-scape is, like its body politic, highly fragmented.

Despite so much of American mythology and iconography that extols purported country values, virtues, and people, the dominance of urban and metropolitan interests has long been an established economic, social, and political fact. As Danbom's (1979) rich and provocative study *The Resisted Revolution* makes abundantly clear, even rural reform efforts, such as the turn-of-the-century Country Life Movement, were essentially arguing only about the terms of surrender. As Cronon (1991) forcefully argues in *Nature's Metropolis,* most of the nation's development has been based on direct linkages between rural and urban and between nature and society. Finding a niche for rural people and their communities in the unfolding urban and industrial regime became the imperative for rural leaders and interests. And rural America has been adjusting itself ever since in an effort to meet the priorities and expectations of the nation as articulated by metropolitan political and economic elites (Stauber, Chapter 8). Central among these has been the goal of establishing a stable, highly predictable, generally healthful, and cheap food-supply system that would meet the needs of a burgeoning urban-industrial and now postindustrial population. It was beside the point that out of this would emerge an ever-dwindling and aging rural population characterized by increasing pau-

91

perization, the continual out-migration of the young and productive, and fewer options for those remaining in the countryside. Food had to be produced, timber had to be cut, minerals had to be mined, fish had to be caught, water had to be diverted, and industrial wastes had to be buried.

Among all types of rural people, farmers were subject to the greatest degree of public policy scrutiny and experimentation. The system that evolved to address the problems of farmers remains one in which meeting urban consumers' needs and generating massive export trade earnings still are the priorities. Despite our national romance with the idealized small, independent farm family living and working in the country on its own land, public policy has encouraged greater market specialization, larger rather than smaller family-based units of production, the commercialization of innovation and agricultural science, the intensification of capitalization—and with it the greater use of chemicals and expensive machinery—and the general commodification of rural life. How could it be any different given these national goals?

We believe, however, that the potential exists for genuine change, that policy is in flux, and that some old assumptions, paradigms, and stereotypes holding back change are being questioned as Americans reassess the nature of their food system, the health and integrity of the environment, and the nature of community in national life (Drabenstott and Barkema 1995). Furthermore, we see that the purported division between urban and rural can be challenged and that new connections between the city and country are emerging, largely because of the growing importance of metropolitan agriculture. Metropolitan agriculture holds the promise of bringing urban and rural people closer together, initially around the marketing of produce, meats, grains, and other commodities both in growing rural/urban fringe areas and in traditionally unserved inner-city communities (Nettleton 1996). Ours is a vision of a new American agriculture, built on an entrepreneurialism rooted in both community and environmental responsibility, that promotes producer-consumer cooperation, a shared commitment to a negotiated landscape combining elements of the city and the country, and community-based food-security systems.

The New Metropolitan Agriculture

Simplistically called "hobby farming," metropolitan agriculture is an increasingly effective form of local economic development and landscape preservation and is a potential bridge between urban and rural communities and people. Because its growth has been incremental across a vast amount of territory and because Americans have a particular sociocultural image of what farming is and what it should look like, metropolitan agriculture has largely been invisible. The emergence of this form of agriculture is the result of economic opportunity and the landscape and lifestyle preferences of rural producers, coupled with the urban consumers' desire for fresh, plentiful, safe, and healthy commodities. Together they are forging a viable and highly entrepreneurial agriculture.

Nationally, approximately one-third of all farms are located within Metropolitan Statistical Areas. In the Northeast, metropolitan farms rise to one-half the total, and in the Pacific Northwest, their share is approximately two-thirds (Heimlich 1989a). Metropolitan area farms produce more than two-thirds of all fruits and vegetables and three-quarters of greenhouse, nursery, and horticultural crops. A recently published American Farmland Trust study showed that the country's 1,549 "urban-influenced counties"—a less-restrictive definition than metropolitan counties, which are defined purely by census criteria—produced 87 percent of all fruits and nuts and 86 percent of all vegetables. Even for products raised on a larger scale, such counties are highly productive. These farms produce 47 percent of all grain, 45 percent of all livestock and poultry, and 79 percent of all dairy products nationally (American Farmland Trust 1994).

Until the 1960s and 1970s farming on the rural/urban fringe, which realistically may extend outward 10 to 40 miles from established urban and suburban centers and areas, was generally understood to be something of a relic agriculture. At that time older farms witnessed some significant changes in their neighborhoods. They were now becoming interspersed with new housing and commercial developments as exurban growth became an important and pervasive national demographic and economic trend. Advances in transportation and communications technologies, the deconcentration of employment and services, and preferences for large-lot residences in rural or semirural locations combined to create a new land-use dynamic throughout the nation. Nowhere was this more observable than in the highly urbanized Northeast.

Although the spread of new residential developments into traditional farming territory brought with it problems, many of which can be subsumed under the broad rubric of right-to-farm issues (Lapping and Leutwiler 1987), both the priced and nonpriced goods and services that farmers have long produced—fresh and plentiful supplies of food, open space, landscape and scenic quality, wildlife habitat, and water and air quality—found greater recognition and appreciation. In a real sense their value increased. All these amenities helped to create an overall quality of life that allowed fringe areas to retain people while attracting new residents. Along with some old-timers who remained in established farming communities that historically served large urban centers, new farmers moved into the fringe to take advantage of emerging market opportunities and lifestyle options. Because land settlement in rural/urban fringe counties is generally uneven and often accurately described as "leapfrog," large portions of such regions, which can be vast in themselves, remain essentially rural even though they are close to established cities and suburbs.

Nationally, the number of farms declined and farm sales stagnated in metropolitan areas during the 1960s, but both increased in the 1970s and 1980s. Growth in the number of farms and agriculturally related sales was accompanied by an increase in cropland, even though the total amount of land in farms continued to decline. This fact has been interpreted as a sign of the intensification of metropolitan agriculture during this period.

The Northeast Region: A Case Study

Our analyses show that this pattern continued throughout the 1980s in many metropolitan counties throughout the Northeast (Pfeffer and Lapping 1994, 1995a, 1995b). The total number of farms and amount of farmland in the rural/urban fringe remained surprisingly stable throughout the decade. These trends have led several observers to conclude that agriculture is a dynamic sector of the economy of rural/urban fringe areas and that it is becoming far too important to be ignored (Heimlich 1989b). Agriculture has, then, flourished in portions of metropolitan areas. Compared with their nonmetropolitan counterparts, metropolitan farms tend to specialize more in high-value commodities and sell more commodities directly to consumers. Often, their production is geared to the needs of nearby ethnic communities. We have noted a real growth in the number of farms whose lambing operations, for example, cater to the needs of residents of Middle Eastern origin. Similarly, more vegetable producers have introduced an array of Oriental vegetables for Asian communities and others who enjoy that cuisine. Metropolitan farms also tend to be smaller and more intensive in their use of resources. They are, in short, more intimately connected with their markets and highly flexible in responding rapidly to changes in the marketplace. They generally do more with less. Metropolitan farmers tend to follow what Smith (1987) has called a "value mode of agriculture," which combines these aspects of agriculture.

Heimlich, perhaps the leading student of metropolitan agriculture, interprets this to mean that farmers adapt over time to take better advantage of new opportunities available within the metropolitan setting (Heimlich and Barnard 1992). This observation has also been made by others (e.g., Johnston and Bryant 1987).

This process of adaptation led to some significant structural changes in Northeastern metropolitan farming in the 1980s. Between 1978 and 1987 the number of smaller farms operating less than 100 acres increased, while the number of middle-sized farms of 100 to 500 acres declined substantially. The types of farms changed also. The largest declines were in the number of farms specializing in dairy and poultry production, and the greatest increases were in farms that sold mostly other animal products, mainly horse farms. In this respect a significant segment of the region's metropolitan agriculture shifted from farm production to leisure pursuits. Nevertheless, such a change can have a positive effect by preserving open space and farmsteads. In contrast to marked changes in livestock production, the numbers of farms earning most of their income from field crops, horticultural specialties, other livestock, and vegetables were stable.

We have identified three distinct types of metropolitan agricultural activity and land use: areas that have achieved a degree of stability in type of farming; areas that are seeing substantial changes in income source but show overall stability in the number of farms and amount of land in agricultural use; and areas witnessing very substantial changes in either farm structure or the type of farm production. The types of changes, with their associated geographic locales in the Northeast, are outlined in Table 7.1, which also shows specific types of farming dominant within each of the three general patterns identified. These areas are typified by some general characteristics:

1. *Areas of stability:* In areas of stability the main source of farm income did not change markedly in the 1980s. These areas are distinguished by a higher rate of farm loss than for the region as a whole, while losses of farmland and gains in population were about the same as for the region overall. One exception is areas dominated by dairy farming, which experienced much slower population growth than any other type of area in the region.

2. *Areas of farmland stability and change:* These areas showed strong declines in some principal income sources but stability in others. They experienced higher rates of farmland losses than the region as a whole (e.g., an 11.9 percent farmland loss between 1978 and 1987 compared with an 8.9 percent decline for the region as a whole). Counties where diversified crop farming is the main income source had the highest

Table 7.1. Areal typology of agriculture in the metropolitan Northeast

Area Type	Agricultural Land Use Characteristics	Geography
1. Stability		
Small-scale ranching	Stability of area dominated by farms of 50 to 99 acres involved in specialized livestock production other than dairy or poultry	Most heavily concentrated in eastern Md. and western Pa., the Baltimore and Pittsburgh metro areas, respectively; also found in N.Y. around Syracuse and Rochester
Dairy farming	Dairy farms operating between 100 and 1000 acres predominate; these areas are very specialized, with noticeable absence of farms less than 100 acres and of fruit and vegetable production	Almost all found in N.Y. (especially) and Pa., but not concentrated in particular metro areas
Extensive grain production	Strong presence of very large (1000+ acres) specialized grain farms	Largest concentration found in northwestern N.Y. around Rochester metro area
Specialized poultry	Strong stable presence of poultry production reinforced by strong recent growth in number of poultry farms	Clearly concentrated in eastern Pa.
2. Stability and change		
Diversified crop farming	Strong declines in largest farms (1000+ acres), with stable presence of general crop and field crop operations; structure reinforced with growth of general crop farms	Found in northern N.J./ southern N.Y.; also in Pa. around Philadelphia metro area
Diversified livestock production	Stable presence of general livestock farms reinforced by growth in such farms; decline in specialized dairy production	Found exclusively in New England, spread throughout without strong attachment to particular metro areas
3. Transition		
Periurban specialty farms	Change characterizes these areas, with strong growth in small-scale (1 to 9 acres) and animal specialty and other nonspecialized livestock, vegetable, and horticultural specialty production; also growth in field crop production to service local livestock	Concentrated in Mass. and N.J. (especially the greater Philadelphia metro area) but also found in Pa. (outside the Philadelphia area) and in Vt.

Table 7.1. (*continued*)

Area Type	Agricultural Land Use Characteristics	Geography
Small-scale fruit	Strong growth in small-scale (10 to 49 acres) fruit production	Central N.J. and Hudson Valley are the main concentrations
Mid-scale extensive grain	Growth in number of grain farms less than 500 acres	Especially important in eastern and western Mass. (Berkshire and Middlesex counties)
Agricultural decline and land consolidation	Strong growth of middle-sized farms (50 to 500 acres) but declines in all types of agricultural commodity production; likely that farm consolidation is part of effort to convert land to commercial or residential uses	Found almost entirely in Baltimore and Washington, D.C., metro areas

rate of farmland loss of any area type in the region. On the other hand, there was almost no change in the number of farms between 1978 and 1987.

3. *Areas of transition:* Change in the main source of farm income defines areas of transition. One type of county is characterized by very high population growth coupled with declines in farms and a steep rate of decline in farmland. These areas are experiencing an exodus from agriculture. Other areas of transition are the locus of agricultural vitality, where strong population growth is coupled with actual increases in farm numbers and farmland losses are lower than in the region as a whole. Agricultural alternatives are surfacing most strongly in these areas of transition.

Nationally and within the Northeast, large parts of the metropolitan agricultural community are actively developing alternative models of farm organization that depart from the highly specialized, government-supported and -regulated agricultural system that for so long has dominated the national food-supply system. The diversification of farm operations and production is geared to specific market niches offering premium price opportunities in rural/urban fringe areas. Producers with direct access to urban and suburban consumers appear to get better prices that help to compensate for greater risks of operating without government price or income support guarantees. A very large fraction of such producers combine farming with other income-generating activities. Some of these are farm-based, such as tourism, animal boarding and training, and estate management. Still, most nonfarm production income is earned in nearby urban communities by either the farmer or other members of the farm family. Table 7.2 illustrates the kinds of products and services that local planners in the Northeast expect metropolitan farmers to continue to provide (Pfeffer and Lapping 1994).

Reality and Promise of Metropolitan Agriculture

Despite these potential and emerging opportunities, several problems continue to threaten farm survival in the metropolitan Northeast and other regions. Policymakers must recognize that metropolitan agriculture creates wealth, renews communities, conserves and protects critical environmental resources, provides many amenities, and serves as a point of connection, a true common ground, for urban and rural people. Effective policies will contain a mix of market-driven incentives and complementary public-sector programs that reflect contemporary social priorities and intergenerational goals. These will seek to integrate land-use and community economic policy in a way that recognizes agriculture as an essential activity in most rural/urban fringe regions across the nation.

Table 7.2. Products and services for which professional planners in the metropolitan Northeast expect increased demand

Food/Feed Crops	Animals/Products	Other Products	Services
Animal feed (hay and small grains)	Fish	Bark mulch	Animal boarding, breeding, and training
	Goat's meat and milk	Fuel wood	
Fresh fruits and vegetables (especially organic)	Horses	Pulp wood	Farm retreats, tours, and vacations
	Lamb	Saw timber	
		Wood	Food delivery (mail or direct) and service
Greens	Local beef and pork	Bedding plants	
Herbs	Organic eggs and poultry	Cut flowers	Hayrides
Maple syrup		Trees	School field trips
Mushrooms	Specialty cheese	Turf/sod	Wine tasting
Table grapes	Veal	Fruit baskets	
Wine	Venison	Jellies	
		Pies	

Source: Authors' survey of 210 professional planners.

The viability of agriculture in rural/urban fringe areas is most immediately threatened by the conversion of farmland to alternative uses. Development pressures directly harm agriculture by increasing property tax burdens on farm operators, reducing the number of farms to below the critical mass necessary to maintain the profitability of essential agricultural support businesses, and making it more difficult for new farmers to enter agriculture because they must bid against developers for increasingly expensive land. Furthermore, sometimes an impermanence syndrome develops, where existing farmers no longer invest in their operations, believing that imminent land conversion will make any investment irrelevant. The result is that these farms often fail (Coughlin and Keene 1981). Large farms are often subdivided into smaller units as a result of development pressures. This leads to a checkerboard distribution of farmlands, with few contiguous fields, making it difficult for farmers to monitor crop growth, move equipment efficiently, and control pest populations effectively.

A variety of strategies to preserve farmland have been developed across the nation. Agricultural zoning (Pennsylvania's Lancaster County, for example), urban growth boundaries to control sprawl (Oregon), combined zoning and income and property tax relief (Wisconsin), development rights transfers (New Jersey), and other techniques have been implemented, all with some effect.

The greatest concentration of agricultural land preservation programs exists in the Northeast, largely because of urbanization, the loss of open space amenities, and the desire to maintain an agricultural industry (Lapping 1979, 1980; Lapping et al. 1989) Some approaches, such as differential farmland assessment, agricultural districts, right-to-farm laws, and the purchase of development rights, have been widely adopted. Differential farmland tax assessment, under which farmland is assessed according to its value in farming, not its development value, is the most common measure, having been adopted in some form in every state in the nation. Agricultural districts exist in New York, Pennsylvania, New Jersey, and Maryland. Purchase of development rights programs are found across the region (Freedgood 1991; Daniels 1991). Right-to-farm laws have been widely adopted. They seek to protect farms from nuisance suits brought by nonfarming neighbors attempting to enjoin certain standard farm practices. The emphasis has been on both open space and farmland preservation. These and other land-conservation and growth-management policies have had some success in protecting farmlands. It is hard to imagine that the recent gains in metropolitan agriculture could have been achieved without differential assessment and other land-conservation measures. A recent evaluation of Pennsylvania's development rights purchase program, for example, concludes that it has "been successful in preserving viable farmland that might otherwise have been sold for development," precisely the kind of land essential for the expansion of metropolitan agriculture (Maynard et al. 1995).

Although new and more imaginative policies will have to be crafted if the land base for metropolitan agriculture is to be preserved, the reality is that state and local farmland preservation activities—especially when combined with other programs to support and promote agriculture—have had solid results. The likely resolution to these and related problems lies in the continuing maturation of a policy framework that understands the necessity to tie land policy more closely to programs that promote the economic viability of agriculture. Retaining farmland will not amount to much if the opportunity to farm is not also secured. Farmland without farmers simply will not do. Additionally, a new set of measures will be required to bring greater coordination between society's infrastructure needs—roads, water and sewage facilities, utilities—and its land-use goals. Such capital investments often act as catalysts for growth and can radically alter local land markets. It will be necessary to control both the location and cost of these investments as they relate to agriculture. This will likely mean that more states will have to follow the lead of Oregon and a few others by defining which areas they wish to see grow and which they wish to preserve and then limit development in the latter by controlling the location and timing of growth-inducing infrastructure.

Various other measures have been enacted to address the specific financial needs of farm enterprises in both the fringe and elsewhere: land trusts, purchase of land conservation easements, relaxation of local zoning requirements for roadside markets and housing for farm laborers, maintenance of buffer zones to protect farmers from conflicts with nonfarming neighbors, and assistance with the development of markets for farm commodities. Programs that provide incentives for the purchase of locally produced foods by public institutions, such as schools, prisons, and nursing homes; the establishment of farmers' markets in cities (often privately managed but sponsored by local agencies, the U.S. Department of Agriculture, or Cooperative Extension); and food origin labels, such as the "Vermont Seal of Approval" or "New Jersey Fresh," all help agriculture economically, in metropolitan areas and elsewhere.

Nongovernmental programs also benefit both the city and farm by forging connections between them. Significant among these has been the sharp rise in community-supported agricultural operations, or

CSAs. CSAs are not necessarily the result of a "program"; they can be initiated by an individual farmer acting alone. The American Farmland Trust recently estimated that approximately 500 CSAs are in operation throughout North America (Berton 1995). A survey by Van En (1995) found about 550 CSAs in operation throughout the United States, with a third of them in the Northeast. CSAs are a mechanism for sharing risk between consumers and farmers. Consumers purchase a share in the farm operation, perhaps also committing a specified amount of labor to be provided on the farm, in exchange for part of the farm's bounty throughout the year. Farmers participating in CSAs reduce their risks from the vagaries of weather and prices and avoid many costs required to market their produce off-farm (Suput 1992; Sherman 1996).

Some suggest that the most significant attribute of CSAs is that they teach people long alienated from agriculture how food is produced and how land and other resources can be managed more wisely (an issue discussed more broadly by Kate Clancy in Chapter 4). To date, most CSAs are run organically, with heavy use of manures, crop rotations, composting, and integrated pest management, and with labor substituted for capital in the farm enterprise.

CSAs also can be important in mediating some conflicts between farmers and nonfarming neighbors over problems such as noise and odors. While almost every state has enacted some form of right-to-farm law that seeks to protect agriculturalists from nuisance suits and local ordinances seeking to prohibit standard agricultural practices, these laws currently protect only already established farmers. But because metropolitan agriculture is so dynamic, some form of protection for new farmers may also be necessary. Because CSAs seek to establish linkages and understanding between farmers and their neighbors, they can act as vehicles for a negotiated landscape, so that all parties will have their interests reflected in the rapidly changing rural/urban fringe environment. In this way, farms and nonagricultural land uses can coexist. It is especially important that nonfarming interests understand that farms create many of the landscape and amenity values that attract people into the rural/urban fringe. Likewise, it is important that farmers understand that the nonfarming public also has legitimate concerns over some externalities of agriculture. CSAs may help everyone to understand better the nature of the challenges facing producers, consumers, and residents in areas of dynamic change.

Much of the future of metropolitan agriculture lies in the rediscovery of the validity of an age-old truism: "Cities aren't problems. . . . They're not places to feel sorry for—they're markets," as Milwaukee

Mayor John Norquist recently asserted (Wiff 1994). But in the end this alone will not suffice. Agricultural literacy must also be renewed so that both producers and consumers understand what is at stake as Americans reform, realign, and modify their food system. Seeing one another as little more than producers and consumers—as part of nothing larger and more substantial than an economic exchange system—fails to capture the rich promise of enhanced urban and country ties and connections. Both types of communities depend on one another to maintain the integrity of the natural systems on which all life depends. A failure to understand this reciprocal relationship and to base farm-city connections and cooperation on it is a folly that leads to ecological and economic disaster. Further recent studies suggest that personal connection—a stronger sense of community that brings farmers and consumers closer together—is an objective of both communities (Lockeretz 1986; Roth 1996; Clancy, Chapter 4). Community and regional renewal promoted by enhanced cooperation based on reciprocity can be achieved as urban people come to understand that agriculture provides much more than food, and as rural people—perhaps led by those living and working on the rural/urban fringe—recognize that their city counterparts are much more than just "another mouth to feed."

Other imaginative policies also will be necessary for metropolitan agriculture to flourish. Paramount among these is a change in mindset, one that sees metropolitan agriculture as a potentially significant form of local and regional economic development. Governments are quick to embrace factories and office parks as signs of economic development and growth. They will often provide extravagant economic incentives—tax benefits, zoning and planning allowances, financing, training, etc.—to influence the location of firms. However, these approaches rarely create new wealth. Rather, they rearrange the location of production as firms seek to take advantage of these "beggar thy neighbor" schemes that communities continue to inflict on one another. Governments seldom see the value of providing incentives to keep or attract farms and farm-based businesses in their jurisdictions. This is especially the case in rural/urban fringe areas, where farming is too often perceived as little more than a lifestyle choice—a hobby at best. The growth and importance of metropolitan agriculture should work against this tendency.

The importance of metropolitan agriculture as an economic activity also extends to its landscape and environmental conservation benefits (Nassauer, Chapter 5; Freedgood, Chapter 6), which are very important in attracting people and businesses into the rural/urban fringe. Local governments can create a climate favorable to agriculture in several ways: by flexible zoning and planning regulations that favor land conservation and the establishment of farm-related enterprises such as roadside stands and marketplaces; by enforcing right-to-farm laws; by cooperating in marketing local produce, especially to local institutions; by sponsoring land trust activities to protect large, undeveloped pieces of farmland; and by supporting other business-related needs of this emerging sector. This will happen only when agriculture in general and metropolitan agriculture more specifically are seen as legitimate and valuable forms of wealth creation, economic development, community renewal, and environmental conservation. In our vision, metropolitan agriculture is a growth industry.

Reference List

American Farmland Trust. 1994. *Farming on the Edge: A New Look at the Importance and Vulnerability of Agriculture Near American Cities.* Washington, D.C.: American Farmland Trust.

Berton, Valarie. 1995. Farm partners: Community-supported agriculture educates consumers, protects farmland. *American Farmland,* Summer: 4–7.

Coughlin, Robert E., and John C. Keene, eds. 1981. *The Protection of Farmland: A Reference Guidebook for State and Local Governments.* Washington, D.C.: U.S. Department of Agriculture, National Agricultural Lands Study.

Cronon, William. 1991. *Nature's Metropolis: Chicago and the Great West.* New York: W.W. Norton & Co.

Danbom, David. 1979. *The Resisted Revolution: Urban America and the Industrialization of Agriculture.* Ames: Iowa State University Press.

Daniels, Thomas L. 1991. The purchase of development rights: Preserving agriculture and open space. *Journal of the American Planning Association* 57 (4): 421–30.

Drabenstott, Mark, and Alan Barkema. 1995. A new vision for agricultural policy. *Economic Review, Federal Reserve Bank of Kansas City* 80 (3): 63–78.

Freedgood, Julia. 1991. PDR programs take root in the Northeast. *Journal of Soil and Water Conservation* 46 (5): 329–441.

Heimlich, Ralph E. 1989a. *Metropolitan Growth and Agriculture: Farming in the City's Shadow.* Report AER-619. Washington, D.C.: Economic Research Service, U.S. Department of Agriculture.

———. 1989b. Metropolitan agriculture: Farming in the city's shadow. *Journal of the American Planning Association* 55 (4): 457–66.

Heimlich, Ralph E., and Charles H. Barnard. 1992. Agricultural adaptation to urbanization: Farm types in Northeast metropolitan areas. *Northeastern Journal of Agricultural and Resource Economics* 21 (1): 50–60.

Johnston, Thomas R. R., and Christopher R. Bryant. 1987. "Agricultural Adaptation: The Prospects for Sustaining Agriculture Near Cities." In *Sustaining*

Agriculture Near Cities, edited by William Lockeretz, 9–22. Ankeny, Iowa: Soil Conservation Society of America.

Lapping, Mark B. 1979. Underpinning for an agricultural land retention strategy. *Journal of Soil and Water Conservation* 34 (3): 124–26.

———. 1980. "Protection Efforts in the Northeastern States." In *Protecting Farmland,* edited by Frederick Steiner and John F. Theilacker, 173–81. Westport, Conn.: AVI Books.

Lapping, Mark B., and Nels R. Leutwiler. 1987. "Agriculture in Conflict: Right-to-Farm Laws and the Peri-Urban Milieu for Farming." In *Sustaining Agriculture Near Cities,* edited by William Lockeretz, 209–18. Ankeny, Iowa: Soil and Water Conservation Society of America.

Lapping, Mark B., Thomas L. Daniels, and John E. Keller. 1989. *Rural Planning and Development in the United States.* New York: Guilford Press.

Lockeretz, William. 1986. Urban consumers' attitudes toward locally grown products. *American Journal of Alternative Agriculture* 1 (2): 83–88.

Maynard, Leigh J., Timothy W. Kelsey, and Stanford M. Lembeck. 1995. Pennsylvania's agricultural conservation easement program: The first three years. *Penn State Farm Economics,* November/December, 1–4.

Nettleton, John S. 1996. "Regional Farmers' Market Development as an Employment and Economic Development Strategy." In *Environmental Enhancement through Agriculture,* edited by William Lockeretz, 235–43. Medford, Mass,: School of Nutrition Science and Policy, Tufts University.

Pfeffer, Max, and Mark B. Lapping. 1994. Farmland preservation, development rights, and the theory of the growth machine: The views of planners. *Journal of Rural Studies* 10 (3): 233–48.

———. 1995a. Public and farmer support for purchase of development rights in the metropolitan Northeast. *Journal of Soil and Water Conservation* 50 (1): 30–34.

———. 1995b. "Prospects for a Sustainable Agriculture in the Northeast's Rural/Urban Fringe." In *Research in Rural Sociology and Development: A Research Annual,* edited by Harry Schwarzweller and Thomas Lyson, 67–93. Greenwich, Conn.: JAI Press.

Roth, Cathy. 1996. Net benefits of community supported agriculture to Northeast farmers and consumers. Paper read at the North American Farmers' Direct Marketing Conference, 23–24 February, Saratoga Springs, N.Y.

Sherman, Robin. 1996. Selling the farm: Innovative farms, marketing strategies, and prospects for sustainable agriculture in Massachusetts' Connecticut River Valley. M.A. thesis, Tufts University, Medford, Mass.

Smith, Stewart. 1987. "Farming Near Cities in a Bimodal Agriculture." In *Sustaining Agriculture Near Cities,* edited by William Lockeretz, 77–90. Ankeny, Iowa: Soil Conservation Society of America.

Suput, Dorothy. 1992. Community supported agriculture in Massachusetts: Status, benefits, and barriers. M.A. thesis, Tufts University, Medford, Mass.

Van En, Robyn. 1995. Eating for your community. *In Context* 42 (Fall): 29–31.

Wiff, Judy. 1994. Cities aren't problems—They are markets. *Wisconsin Rural Leader.* March 5.

8

Envisioning a Thriving Rural America through Agriculture

Karl Stauber

A Vision of the Future

Agriculture used to be the focus of much of rural America. Many people, especially urban and suburban people, still think it is. But farming and ranching have been declining players in much of rural America for most of this century. As agriculture has declined in economic significance, we have seen three basic patterns emerge: communities have seen the wealth from agriculture become more concentrated; regions have experienced substantial declines in population and economic vitality; or agriculture has been replaced by growth in other sectors such as manufacturing and tourism or displaced by competing land uses such as suburban development. The hard fact is that as agriculture as an economic and political force has become concentrated in fewer and fewer hands, rural America in general has not benefited.

For example, of the 435 congressional districts in the United States, by 1990 only 17 percent were more than one-half rural. Only 8 percent had at least 5 percent of their total population living on farms. The district with the highest proportion living on farms was Iowa's Fifth District, at 14 percent of the district's total population (Calvin Beale, USDA-ERS, personal communication, 1994). In 1992, for the first time in U.S. history, the majority of votes for president were cast in suburban districts. In 1995, again for the first time in U.S. history, none of the top five leadership positions in the House of Representatives was occupied by a member from an agriculturally dependent or rural district. America is now a suburban nation. The political power base that rural America, and particularly agriculture, held for so long is no more.

Must this trend continue? I think not.

Changes in private-sector initiative and government policy can stim-

105

ulate the reemergence of agriculture as a force for the economic vitality of rural communities. Agriculture will never again be the only force for rural vitality, but it can and should be part of a diverse economic base for many communities.

Governmental policy must be refocused to support increased economic opportunity in agriculture rather than the course of increased concentration that we are currently on. The government's implementation of the 1996 federal farm legislation and the private sector's response are great opportunities for this redirection. If current market forces are left to play out, agriculture will be concentrated in fewer hands, rural poverty will increase in some parts of the United States, and other rural areas will continue their economic and political decline.

Redefining Governmental Rural Development Policy

When the market fails, government has a responsibility to intervene. As outlined below, rural America includes many examples of market failure. Rural America also has many examples of government failures. From the successes and failures of past government efforts, a core set of assumptions that should underlie all government rural development initiatives (Galston and Baehler 1995, 267) include the following:

- No single approach to rural development will be appropriate to all rural communities. The "one size fits all" approach of past national policies must be replaced with efforts that allow for variations in opportunity and problems.
- The for-profit economy—the marketplace—is, and should be, the dominant force in rural development. Whenever possible, we should rely on the marketplace to direct and support rural development. Government should not subsidize people and communities to do what the marketplace can do better.
- Rural communities and the marketplace—not national programs—will determine whether areas prosper or decline. However, there are occasions when the federal government has the responsibility to create equal opportunity. Not all areas have equal access to political and economic power. Historical and current patterns of discrimination are an important cause of poverty. This is not acceptable. Furthermore, competition and the marketplace have rarely overcome such problems by themselves. Therefore, government actions are needed to ensure that traditionally disadvantaged communities and individuals have the opportunity to compete.

- In some parts of rural America, *community* development must precede and then overlap with *economic* development. If communities are to prosper, *all* people must have access to certain basic services such as water, health care, education, and transportation. Without such essential services, a community will be constrained in its ability to participate in the marketplace.
- Infrastructure alone will not produce community development. Leadership development, the creation of locally controlled institutions, and experience in managing and owning assets also are critical elements of community development, but none of these alone will be adequate.
- Federal finances will continue to be constrained by the budget deficit. Therefore, there will not be adequate federal resources to meet all the needs of rural America. Federal resources should be targeted to places where opportunity is declining or stagnant and where the marketplace will not be able to provide adequate economic activity to ensure opportunity for all. This includes approximately five hundred persistent poverty counties, such as in the Mississippi Delta and Appalachia, and seven hundred counties experiencing shrinking population and job opportunities, such as in the Great Plains.
- The future of rural areas is and will be tied to the viability of urban areas. Although rural Americans may wish to separate and isolate themselves from their urban cousins, they cannot have both separation *and* opportunity. In fact, linkages to urban America are critical to rural opportunities (Lapping and Pfeffer, Chapter 7).

If rural development policy is to succeed, it must explicitly focus on reducing the widening gap between rural America and the rest of the nation. National economic prosperity has rarely trickled down to most rural communities in recent decades. Aggressive federal action is needed to change this fact.

Where and What is Rural America?

Rural America is both a place and a concept. Rural America has multiple images and places. In popular culture its image ranges from *The Bridges of Madison County* to *The Last Picture Show,* from "Picket Fences" to "Northern Exposure," from the home of the Michigan Militia to Bill Clinton's hometown of Hope, Arkansas.

Except for a few television programs or the stories of their grandparents, the view Americans have of rural America is ill informed. Most Americans have never lived in rural America and most see rural communities as places of plenty, free of crime and stress, with significant economic opportunities. A 1992 survey by the Roper organization showed a widely held perception about rural America:

> Americans continue to have an enduring admiration for rural Americans. Rural Americans continue to be thought of as family oriented, friendly, honest, responsible, religious, and less stressed than their urban counterparts. Another enduring impression is that urban and suburban problems do not affect rural America or constitute an important threat. . . . The public does not recognize or think about the poverty and social problems prevalent in rural America. Rural America itself may have changed, but the way in which most Americans view it has not. In the eyes of most Americans, rural America has an embarrassment of riches, not problems (Bonnett 1993).

In reality, rural America is complex and characterized by a variety of opportunities and needs. There is a growing disparity in economic health between rural and urban America. During the 1980s rural growth slowed sharply, even as urban areas boomed. For the United States as a whole, per capita rural incomes grew only sixty cents for every dollar of metro income growth (Drabenstott 1993). In aggregate, rural areas have higher rates of poverty, unemployment, and mortality; lower levels of educational attainment, employment skills, and vocational training; and more limited access to health care, social services, public water systems, and modern telecommunications (Flora and Christenson 1991, 333).

Economic opportunities are expanding in areas next to urban communities, rural areas with scenic beauty or recreational amenities, and areas with other competitive advantages. But economic opportunities are stagnant in areas farthest from population centers, and some parts of rural America are in serious economic decline. The declining areas include regions where past and continuing patterns of racism have produced uneven distributions of education, access to capital, and other precursors of growth.

One thing is clear: "Rural America is no longer agricultural America. During most of their history, farming was the foundation of rural economies. But now only about one-fourth of rural counties, mostly in the Midwest, can be regarded as truly dependent on agriculture. And these areas account for less than seven percent of the U.S. non-metropolitan population" (Corporation for Enterprise Development 1993).

This confusion about the nature and place of rural America is important because it constrains our ability to create programs and policies that encourage rural economic development. In a time of constrained or declining government resources, targeting becomes critical. Poor definitions and understanding lead to poor policy. Poor policy can hasten the decline of areas of greatest need while assisting areas that will experience increased opportunity from the marketplace.

Past and Current Rural Development Policy

Current federal, state, and tribal rural development policies are confusing, reflecting a variety of historical periods. At first, almost all government development policy also was rural policy because America was a country defined by the "frontier." Government policy focused on occupying rural America. But around one hundred years ago we completed the "settlement" phase (this label ignores the well-established Native American civilizations) and shifted to mostly urban industrialization. Rural America became the "storehouse" from which high-quality people, capital, food, timber, and so forth were efficiently extracted. During this period, many government policies were created to improve the extraction of rural resources and support the modernization of urban America.

Particularly beginning with the Great Depression, other policies focused on the problems of rural poverty. The poverty alleviation or "poor house" policies of the 1930s attempted to help the displaced farm families survive by providing food and temporary shelter. In the 1960s, additional initiatives added long-term housing for low-income people and targeted infrastructure development.

Most recently, we have seen the accelerated development of the "playground" period, with parts of rural America seen primarily as places for recreation, not development. Under this approach, some rural areas have become places where urban and suburban Americans go for rest, spiritual rejuvenation, holiday, and retirement. This has been supported by an array of government policies, such as cheap energy, extensive highway systems, and changing use patterns in state and tribal lands and national forests and rangelands.

Today, we have multiple sets of government rural development policies—frontier, storehouse, poor house, and playground—operating simultaneously. This produces a "layer cake" effect, in which policies and programs are stacked up over time. Rarely does a new initiative eliminate the previous layer, thereby producing confusing sets of regulations, authorities, and programs. The resulting waste and duplication must be managed by individuals, businesses, nonprofit groups, and local governments.

In the twentieth century, rural development policy has focused on increasing farm income, recruiting manufacturing jobs to use surplus labor and reduce population decline, and providing infrastructure-based amenities, such as electricity, water, and highways (Carlisle and Rist 1993, 12). These efforts have been highly successful for *some* rural people and communities. We should declare success for these strategies and move on to new approaches. If we stay with these no longer relevant strategies, we will reinforce the difficulties so starkly identified by Flora and Christenson (1991, 333): "In many ways, the social, economic, and cultural problems that many rural Americans face are similar to those faced in less developed countries: . . . a high level of poverty, isolation from economic opportunities, limited access to credit and technical assistance, government incapacity, an inadequate infrastructure, garbage disposal, toxic waste, substandard schools, health care, isolation of certain groups (especially the elderly), and discrimination. . . . Rural America is becoming a separate impoverished 'outland' within American society."

Current government rural development policy is largely bankrupt. It is based on the political power bases of the past, not the opportunities of the future. Commodity organizations, groups that build rural housing and infrastructure, and land-grant universities all are committed to the maintenance of the status quo—a status quo in which those with resources fight over how to distribute the declining base rather than what makes the most sense for rural areas.

Despite the steady deterioration of rural economic conditions, there has been no coherent or effective policy response. Why? According to agricultural economist James Bonnen (1992, 192): "Partly, the failure is due to the lack of well-integrated rural community and political institutions . . . to provide a voice for rural society. It is also due to an agrarian fundamentalism, which, whatever its original validity, is today a myth promoted by cynical farm interests (even including advocates of reformed or 'alternative' agriculture) and by the media."

As long as the vision of the future of rural America is dominated by those who want to hold on to old programs and resource allocations, rural communities will continue to suffer. If rural communities and people are to move beyond the whims of the marketplace in deciding who does well, new government policies and ways of operating must be developed.

There is little persuasive evidence that any major government rural development programs were developed within the framework of any strategic context. As one U.S. Department of Agriculture internal report observed in the late 1980s, "Rural policy at all levels of government consists of a collection of programs that, however useful individually, does not add up to a coherent or consistent strategy to achieve any well-understood goals" (U.S. Department of Agriculture 1989). Little has changed in the intervening years.

Governments must make a clean break with rural development policies that no longer are relevant. Government rural development policy should focus on assisting targeted communities in becoming more competitive in the international marketplace in a way that benefits all citizens, not just those with economic means. Of all the federal initiatives of the last thirty years, the Enterprise Zone/Empowerment Community has come the closest in providing the needed clean break with the past. Unfortunately, the battles over budget cutting and the long-term political power of traditional interests such as commodity groups and rural development organizations will make expansion of these efforts very difficult. Regardless, existing rural programs should be reoriented to focus on the five hundred poorest and seven hundred declining rural counties.

Reenvisioning Federal Policy to Support a Thriving Rural America

Agriculture represents a fundamental linkage between urban and rural Americans. Agriculture is also important in many of the five hundred poorest and seven hundred declining rural counties. Jane Jacobs and other leading rural development scholars have suggested that rural areas with the strongest links to urban areas on average have the most economic vitality (Galston and Baehler 1995, 267). Dan Kemmis, the former mayor of Missoula, Montana, supported the potential of intentional linkages when he said: "As rural life is more severely threatened by international markets, technological dislocations, and corporate domination, it may be time for a reassessment of the relationship between cities and their rural environs. Maybe neither towns nor farms can thrive in the way they would prefer until they turn their attention more directly to each other, realizing that they are mutually complementary parts of the enterprise of inhabiting a particular place" (quoted in Galston and Baehler 1995, 267).

By building on this linkage, public policy and local communities can improve rural prosperity. Agriculture may offer the greatest potential for rural development for many isolated areas that do not have much potential in services and manufacturing. Redirecting current investments from commodity-focused agriculture to development-focused agriculture will be critical to the redevelopment of many rural areas. Agriculture can again become *a* means, although not *the* means, of creating economic opportunities in rural areas.

Although agriculture was the primary focus of rural economic development for much of American history, this will never again be the case. However, if we shift from seeing farming as the primary focus to being part of an integrated, multisectoral approach to creating rural opportunity, farming has great potential as a rural development strategy. Agriculture offers many different approaches to increasing the economic vitality of rural areas; only two will be explored here: buy local initiatives and agricultural industrial processing. Although these approaches are very different, they share several critical characteristics. Both diverge from traditional approaches to rural development through agriculture. They do not rely on traditional agriculture supported by commodity payments or on cheaper production methods for individual farmers. They also do not depend on stealing jobs and small cheap production processes from other parts of the United States (Carlisle and Rist 1993, 14–15).

These two strategies suggest new ways of approaching rural development. They are illustrations of what is possible. The first, buy local, acknowledges that large metropolitan areas drive economic activity. The second, agricultural industrial processing, attempts to make limited scale and low density into advantages. These strategies are illustrations of the different types of approaches that must be developed if more rural areas of the United States are to thrive.

Strategy One: Grow Local, Sell Local=Buy Local

Most of the crops grown by American farmers are not eaten directly by humans. Instead, most of the corn you see as you drive on the rural interstate is used to feed animals. The soybeans that occupy millions of acres in the United States are indigestible by humans until they are processed. But once they are processed, they become industrial feedstocks or animal feed. Only in a few parts of America—such as the Central Valley of California, portions of Florida, and the orchard regions of Michigan, Oregon, Washington, New York, Ohio, and Virginia—is most farmland in crops for the direct consumption of humans.

It wasn't always this way. Fruit and vegetable production used to be spread throughout America. In 1900 Iowa produced more than thirty commercial vegetable crops that were canned or consumed fresh. Today, it produces hardly any such crops on a commercial scale. The same is true of many other parts of the country.

The buy local approach is the antithesis of what has happened in mainstream agriculture for the last one hundred years. We have moved from a local system, supplemented by canned or processed food from other regions, to a modern system in which virtually everything is a "commodity." Commodities, by definition, have lost their specific identity. They are undifferentiated products—one bushel of corn is the same as any other (except for small differences among grades). It does not matter where the corn came from; price is the only critical variable. One reason American food is so inexpensive is that it is so commodified.

A buy local approach to rural development does not mean that American consumers must return to the root cellar as the source of winter vegetables and fruit. But there is evidence that Americans, particularly affluent residents of major cities and their suburbs, are developing a renewed interest in fresh, locally produced fruits and vegetables (Lapping and Pfeffer, Chapter 7). The rebirth of local farmers' markets and the rapid expansion of community-supported agriculture, where consumers share the farmer's risk in return for a say in what is grown and how, are indications of the renewed interest (Clancy, Chapter 4).

Even in Iowa, in the heart of America's traditional corn and soybean production region, buy local agriculture is growing. Although the state lost 34 percent of its farming population in the 1980s and another four thousand corn, soybean, and beef producers between 1990 and 1993, its horticultural production, including fresh fruits and vegetables, increased from $61 to $89 million between 1987 and 1989. It is the fastest growing segment of Iowa's agriculture, according to Dale Cochran, the state's Secretary of Agriculture (Maurer 1995).

Buy local is increasing without significant government support. As suggested above, it is increasing because of market signals and demand. However, if we want buy local to benefit economically disadvantaged regions, government intervention is required.

Buy local represents the greatest opportunity for parts of the five hundred poorest rural counties. To be successful, a targeted buy local rural development strategy must take advantage of proximity to a large metropolitan area, the primary engine of economic activity. Rather than providing commodity payments or production subsidies to the more affluent farmers and ranchers, national, state, and local governments should invest in the creation of new and expanded local, fresh market systems designed to aid rural low-income people in producing, marketing, and processing agricultural products. A significant portion of the resources of publicly supported institutions such as Cooperative Extension and state agricultural experiment stations should be redirected to creating new economic opportunities in the five hundred poorest rural counties through a new approach to commercial agriculture.

Strategy Two: Agricultural Industrial Processing

What are we to do with the approximately seven hundred declining rural counties, which are mostly in the center of the United States? Are we to write off these areas as market failures? If their political power declines, what will be their basis of government support? Are we to accept the "Buffalo Commons" proposed by some Eastern academics? Can rural development use technology in a way that makes the small scale and low density of rural areas into advantages?

Technology has changed the direction and intensity of American agriculture on several occasions. The introduction of irrigation in the very dry "American Desert" of the Great Plains, the modern tractor, the mass-produced moldboard plow, and the cotton gin are but a few examples of technologies that have fundamentally changed farming in various regions of the nation. Many of these technologies have had the unintended consequence of reducing the free, voluntary labor that can be gainfully employed in production agriculture.

However, it is possible for technology to increase on-farm or near-farm opportunities. To date, technology usually has been biased in favor of larger farms. Some have looked to "value-added" activities as the source of more rural opportunity, but the record is not encouraging. Value-added activities add value to locally produced or extracted natural resource commodities before they leave the rural community, such as by turning logs into furniture or corn into ethanol fuel. The USDA Economic Research Service has conducted several surveys of value-added processing. These studies show that rural areas are attractive to corporations, especially for food processing. Overall, employment in food processing declined in the 1980s, but it increased in rural areas. Although this rural expansion has been positive, it is unlikely to continue in the 1990s. Overcapacity in the food-processing sector, combined with increased concentration, is likely to spell declines for many rural areas that depend on processing jobs (Brown and Petrulis 1993). Although these industries have provided some benefits to their immediate rural areas, most often they produce only small profits for their investors and create few good jobs for rural communities.

But new technologies are emerging that may help to reverse this trend. No single strategy will return agriculture to its former level of employment, particularly in the most rural areas. The USDA Agricultural Research Service and the private sector have begun to create new technologies that use supercritical fluid extraction (SCE) to produce new types of plastics and other organic compounds. SCE puts carbon dioxide (or another gas) under great pressure and high temperature. The gas, which is contained in a special vessel, can extract compounds from corn, wheat, or other farm products. These compounds can be used as feedstock for a host of other specialized, high-value products. SCE produces no environmentally damaging side-products, as do many large industrial facilities.

The best-known examples of SCE currently in use are "natural" decaffeination of coffee (*Chemical Engineering* 1995) and flavor extraction from hops and spices (*R&D Magazine* 1995). The USDA's National Center for Agricultural Utilization Research has developed technology for removing fat and cholesterol from meat, whereas other researchers have created similar applications for milk, eggs, and other food products (Cooke 1995). The USDA and other agencies also have used SCE for testing water quality and removing harmful chemicals from contaminated soils (Bosisio 1989).

These microscale industrial processing facilities appear to be most efficient when they are close to the raw material, such as corn, and on a moderate scale. Beginning–scale facilities can be created for around $250,000 to $500,000 per machine. They require highly skilled technical staffs to operate them. The agricultural industrial processing facilities appear to have a scale limit; the physical and chemical activity that occurs within each SCE vessel only works in small volumes (50 to 100 liters). This will keep the agricultural processing facilities small and focused on producing high-value specialty organic compounds. SCE could create a demand for specialty crops that contain specific compounds, and it could produce high-paying rural employment opportunities for highly educated entrepreneurs. Although the capital costs to enter this new sector appear low, the risk will be high until the market is well established. Environmental regulations that limit more traditional extraction techniques may hasten the development of SCE-based rural businesses. Given the risks inherent in any new start-up industry, rural cooperatives may be a logical base for these new efforts.

It is hard to predict the potential of SCE and related microscale agricultural industrial processing (MAIP). Traditional organic extraction is worth billions of dollars per year, but it creates many environmental problems. MAIP could replace the "high end" of traditional techniques, making products that are widely used in drugs, cosmetics, and specialty plastics. SCE technologies also can be used to produce food products.

The Philadelphia laboratory of the Agricultural Research Service is using SCE to turn milk into an imitation fish product highly valued in Asia. This fake fish is highly digestible, reduces pressure on depleted ocean fish stocks, and creates a new market for raw milk.

SCE is but one example of how an aggressive federal, state, and private research program could produce a variety of new rural enterprises. Similar efforts are already under way through the Agricultural Utilization Research Institute in Minnesota and through joint public-private ventures in Florida and North Carolina. Rural communities, especially the seven hundred declining counties, can produce inexpensive feedstocks and with the right research and technology investments turn grain, milk, and other raw farm products into high-value products. However, this will require a long-term investment strategy from government and the private sector.

Conclusion

Agriculture and rural America once were largely inseparable: If farmers and ranchers did well, so did most rural communities. For many rural parts of the United States, it has been more than one hundred years since this was so. However, it is possible to envision a future rural America where new versions of production agriculture form new types of urban-rural linkages. This new vision of rural opportunity moves away from the strategies of the last one hundred years, which focused public resources on reducing the cost of production rather than increasing the scope of opportunity. If rural America is to turn its traditional skills in production agriculture into renewed competitive advantage, it must "decommodify" some of its activities.

I have suggested two ways to start this process of replacing commodity-oriented agriculture with new, market-based urban-rural linkages. These linkages are critical to the future economic viability of rural communities. The first strategy is most appropriate to rural areas near large cities, many of which are in areas of persistent poverty. The second strategy is designed to meet the special needs of the most isolated, declining rural areas.

No strategy will work for all rural communities. However, agriculture focused on development could be the basis for multiple strategies that will benefit many rural areas.

Reference List

Bonnen, James. 1992. Why is there no coherent U.S. rural policy? *Policy Studies Journal* 20 (2): 190–201.

Bonnett, Thomas W. 1993. *Strategies for Rural Competitiveness: Policy Options for State Governments.* Washington, D.C.: Council of Governors' Policy Advisors.

Bosisio, M. 1989. Pinpointing pollutants with soil samples. *Agricultural Research* 37 (8): 11.

Brown, Dennis M., and Mindy Petrulis. 1993. *Value-Added Agriculture as a Growth Strategy.* Agriculture Information Bulletin No. 644-10. Washington, D.C.: Economic Research Service, U.S. Department of Agriculture.

Carlisle, Rick, and Carl Rist. 1993. *Rural Economic Development Trends and Strategies in the United States: Adapting to Change.* Paris: Organization for Economic Cooperation and Development.

Chemical Engineering. 1995. New roles for supercritical fluids. March, 32.

Cooke, L. 1995. Supercritical fluid fat extraction. *Agricultural Research* 43 (3): 18.

Corporation for Enterprise Development. 1993. *Rethinking Rural Development.* Washington, D.C.: Corporation for Enterprise Development.

Drabenstott, Mark. 1993. Prospects for rural prosperity in the 1990s. *Outlook* (March): 900–911.

Flora, Cornelia B., and Christenson, James A. 1991. "A Rural Policy Agenda for the 1990s." In *Rural Policies for the 1990s,* edited by C. B. Flora and J. A. Christenson. Boulder: Westview Press.

Galston, William A., and Karen J. Baehler. 1995. *Rural Development in the United States: Connecting Theory, Practice, and Possibilities.* Washington, D.C.: Island Press.

Maurer, Derek. 1995. Farmers markets promote rural development. *Des Moines Register,* June 27.

R&D Magazine. 1995. 3 supercritical fluid extraction strategies. April, 59.

U.S. Department of Agriculture. 1989. *A Hard Look at USDA's Rural Development Programs: Report of the Rural Revitalization Task Force to the Secretary of Agriculture.* Washington, D.C.: U.S. Department of Agriculture.

9

A Better Life for Farm Workers

Suzanne Vaupel

Immigrant farm workers play a crucial and so far irreplaceable role in U.S. agriculture. However, instead of receiving the gratitude of U.S. consumers, they are often blamed for many of the country's economic and social problems. Most farm workers receive low wages and few benefits for working up to ten hours a day in extremes of heat, dampness, or cold. They work in the second most dangerous occupation in the United States, ranking just behind mining. Dangers include injuries from machinery and heavy equipment, falls from ladders and on wet fields and wet greenhouse floors, and exposure to farm chemicals, dust, and insects.

Some farm workers are plainly exploited—getting paid less than minimum wage and no overtime pay as required by law; working long hours without field sanitation or drinking water; being overcharged for transportation, meals, and cold drinks in the field; and getting fired if they complain. Even cases of peonage have been found in which workers were not allowed to leave a fenced housing compound. Many farm workers live in crowded, unhealthy housing, and some have no housing at all, living instead in cars or packing boxes, or under trees.

The "farm worker problem" is a complex social and economic issue. Farm workers are victims of racial discrimination and discrimination against immigrants, farmers' inability to pass higher costs on to buyers and unwillingness to raise wages, inadequate enforcement of labor protection laws, and exclusion from other legal protections. At the most basic level, they are tied to the unstructured seasonal labor market caused by crop specialization and large, concentrated farms.

The "farm worker problem" has defied solution for more than one

I thank the University of California Sustainable Agriculture Research and Education Program for funding the study of farmers who provide year-round employment and Richard Mines of the U.S. Department of Labor, Theodore Goodwin, Leslie Mitchell, and Thomas Mullins for their helpful comments.

119

hundred years in California. It is more recent but no less serious in other regions of the United States.

A Vision for Farm Workers

It is not difficult to identify the needs of farm workers: higher wages, improved benefits, and better working conditions. The more difficult task is to offer credible solutions that eliminate the root causes of the problems.

Recognizing the need to be realistic in proposing solutions to long-standing problems, I am suggesting a modest vision that would nonetheless have far-reaching consequences for improving the standard of living for farm workers and their families. This vision is of an agricultural system in which workers enjoy wages, benefits, and working conditions that are at least comparable to those of nonagricultural workers and have opportunities for employment outside agriculture if they choose. They will receive a living wage, and their benefits will follow them if they change employers.

Agricultural exemptions and eligibility criteria will no longer exclude farm workers from labor protection laws, and the laws will be adequately enforced. Agricultural workers will have the right to join a union and they can engage in collective bargaining with their employer if a majority vote to do so. They also will have adequate housing, health care, child care, and training in job skills and English.

Finally, farm workers will be accepted as respected members of society and no longer discriminated against because of ethnic or racial background or immigrant status. They will be recognized for the essential role they play in the production of our food supply and no longer be blamed for unemployment and other economic and social problems.

This vision requires many changes. I focus here on three areas: the workplace, the legal system, and the communities where farm workers live. To accomplish these changes, we must thoroughly rethink the problems and develop new solutions. Seventy-five years of public outrage and legislative reform efforts have done little to close the economic gap between agricultural and nonagricultural workers. Unfortunately, resources for labor law enforcement and social programs are decreasing, with little immediate hope for reversal. Yet many other changes occurring could open the door to real solutions, especially a new willingness to look at alternatives. I propose several alternatives that could lead to significant improvement in the lives of farm workers.

This vision is modest because it merely puts farm workers on equal footing with nonagricultural workers. For this to happen, a key factor that differentiates agricultural employment from other work must be addressed: seasonality. I return to this central issue after presenting an overview of the "farm worker problem."

The Importance of Immigrant Farm Workers

Immigrant labor is a well-established part of the agricultural production system in the United States. About a third of all U.S. farms hire farm labor, paying $13 billion in wages annually. Most farm workers are concentrated on large and midsized farms, but 64 percent of farms that hire farm workers have annual gross incomes of less than $100,000 (U.S. Bureau of the Census 1994).

About 60 percent of farm workers in the United States and 92 percent in California are foreign-born (Mines et al. 1993, 11; Rosenberg et al. 1993, 6). Most are from Mexico, with a few from other Latin American countries, Asia, and the Caribbean. The hired farm workforce is concentrated most heavily in the production of fruits, vegetables, and nursery crops. About 40 percent of all farm wages are spent in California, Florida, and Texas, where these crops are produced on large, specialized operations. However, immigrant farm workers are important and irreplaceable in the production of many crops in other states. Workers from Mexico have migrated to the western half of the United States and Florida for more than two generations. More recently, they have migrated to the eastern half of the United States.

Farm Workers in American Culture

Most American consumers know little about how food is produced and even less about the farm workers who plant, hoe, spray, thin, and harvest it. To the public, the foreign accents and low incomes of farm workers are usually associated with the problem of illegal immigration. Immigrants are accused of taking jobs from U.S. workers, living off welfare and other benefits paid for by U.S. citizens, and causing overcrowded schools and increases in crime. Some of this frenzy is stirred up by politicians seeking a scapegoat for economic and social problems. Anti-immigrant legislation and citizen initiatives have taken many forms: the passage of English-only laws, withholding of services such as public schooling and medical treatment, and ending of affirmative action programs. Recent federal legislation will deny illegal and legal immigrants access to food stamps and aid for the aged, blind, and disabled and gives states the option of denying them access to Medicaid and aid to families. Immigration reform efforts have tried to close the borders to illegal immigration and further restrict legal immigration into the United States.

Immigrant farm workers do not fit the stereotype held by most people. Quarterly interviews with more than twenty-five hundred workers in agricultural crops conducted over a three-year period under a mandate of the 1986 Immigration Reform and Control Act showed that 88 percent are legally authorized to work in the United States. Half of all agricultural workers and 62 percent of foreign-born agricultural workers have family incomes lower than the poverty level. Undocumented workers have the lowest incomes, with a median annual income between $2,500 and $5,000. Only 2 percent of undocumented agricultural workers live in families that receive needs-based social services, and in these cases the actual recipient of services may or may not have legal status (Mines et al. 1991).

Farm Workers and Public Policy

Competing policy goals have shaped farm workers' lives for more than seventy-five years. These include the following:

- The goal of reducing the numbers of immigrants, both illegal and legal
- Society's interest in improving the skills and education of farm workers so that they can move into better-paying (nonagricultural) jobs
- The goal of farmers to maintain an adequate labor supply

The Immigration Reform and Control Act of 1986 (IRCA) and recent immigration bills passed by Congress are only the latest in a long series of laws restricting the number of immigrants to the United States. Restrictive legislation passed in 1921 and amended in 1924 limited immigration from Europe and Asia; however, agricultural interests were successful in winning and maintaining exemptions for Mexican immigrants (Fuller 1991, 46–47). Proposals to extend restrictions to immigrants from Mexico were defeated in 1926, 1928, and 1930. During the IRCA debate of the 1980s, agricultural interests led by California again were successful in maintaining various options for continued immigration of Mexican agricultural workers.

The arguments of agricultural employers in favor of liberal immigration from Mexico have been based on agriculture's dependence on Mexican laborers. In 1926, a representative of California agriculture testified before the House Committee on Immigration and Naturalization: "We, gentleman, are just as anxious as you are not to build the civilization of California or any other western district upon a Mexican foundation. We take [the Mexican laborer] because there is nothing else available. We have gone east, west, north, and south and he is the only man-power available to us" (cited in Fuller 1991, 45).

In the IRCA debate of the 1980s, California agricultural employers argued that special programs were necessary to give them time to adjust to a smaller but fully legal workforce. Since the passage of IRCA did not slow illegal immigration, that adjustment was never made.

Like anti-immigration legislation, the social interest in improving jobs of farm workers also works counter to the interests of agricultural employers to maintain an adequate workforce. The primary focus of federal and state employment and training efforts has been to move workers into nonagricultural jobs. Without changes in the basic agricultural system, however, employers simply hire new workers to replace those who leave.

Thus, while two strands of U.S. policy would reduce immigration and move farm workers out of agricultural jobs, the influence of agricultural employers has kept both goals from being attained. A 1940 analysis concluded that public policy regarding agricultural labor was largely a policy of accommodating labor supply to demand (Fuller 1940). In 1984 another analysis found that the same policy had continued to prevail with few exceptions (Mamer 1984, 291).

The Basic Problem: Seasonality

The need for many short-term, seasonal workers developed concurrently with an increase in crop specialization and large, concentrated farming. It was apparent that intensive crop specialization required a fluid labor force that was available when needed but would disappear when the harvest was over. In California, this need was filled by Chinese workers until the 1880s, then Japanese, Punjabi, and Filipino workers, followed by Mexican workers after World War I. The increase of Mexican workers was interrupted only by the Dust Bowl refugees from Oklahoma and Arkansas who worked in California fields from the 1930s to the outbreak of World War II. Since World War II, agriculture throughout the United States has become increasingly dependent on Mexican workers. The problems connected with dependency on foreign workers were well articulated very early, as in this 1920 letter to the California State Board of Control:

> If California is to go on with her agriculture as now organized, she must continue to constantly recruit a supply of labor able and willing to do the hand work necessary to the harvest of many fruits, [vegetables,

and melons]. Either the supply [of labor] must be kept up or a reorganization in our scheme of production is bound to be necessary. . . . There still exists . . . a question as to what the remedy should be. Are we not better off to reorganize on the basis of what we have and to quit fostering industries whose existence depends upon the constant recruiting of such people as Mexicans, . . . or will the economic advantages of a continuation of this sort of thing more than offset the rather evident social disadvantages? It is an important question and upon its correct answering depends the future of our agriculture in many of its important phases (cited in Fuller 1991, 54).

The seasonal agricultural labor market has all the characteristics of undesirable secondary labor markets: low wages and fringe benefits; poor working conditions; high turnover; little opportunity for advancement; little training on the job; and the absence of work rules, seniority systems, formal grievance procedures, and job evaluation plans. Jobs in a secondary labor market are distinguished by their instability and lack of attachment between workers and their employment. Conditions for other seasonal and short-term workers, such as longshoremen and construction workers, improved significantly after the workforce was stabilized through hiring halls. Attempts by the United Farm Workers to use hiring halls to stabilize agricultural work in the 1970s were unsuccessful.

Fruit and vegetable production typically requires many seasonal workers for harvesting, weeding, thinning, and pruning. These jobs last from two weeks to two months. On a small farm, the job may be completed in less than a day. Seasonal agricultural workers are constantly piecing together short jobs with different employers. Only a small core of agricultural workers have regular jobs that last most of the year. These jobs tend to be supervisory or "skilled" positions, such as mechanics, tractor drivers, machine operators, and irrigators.

Labor intermediaries have developed to connect workers with short-term seasonal jobs. Farm labor contractors (FLCs) provide services to farmers that include recruiting, supervising, and paying workers; handling payroll taxes; and sometimes providing transportation, housing, and meals. The wages and benefits paid by FLCs are usually lower than those paid by farmers. In 1983, for example, average annual wages for workers employed by FLCs in California were only 58 percent of the statewide average paid by farmers in the state (Vaupel and Martin 1986). Nationally, only 8 percent of FLC workers receive health benefits, compared with 28 percent of all agricultural workers (Mines et al. 1993, 47).

Agriculture's heavy dependence on large numbers of short-term seasonal workers is shown clearly by one estimate that the 4.5 million seasonal hirings in 1984 equated to fewer than 350,000 full-time, year-round workers (Fritsch 1984, 65). In 1992, 2.9 million farm workers (76 percent of the total) worked fewer than 150 days (U.S. Bureau of the Census 1994). As long as the workforce is predominantly seasonal, there are no real solutions to the problems of farm workers.

Year-Round Employment: An Alternative to Seasonality

Over the years, some growers have successfully structured their agricultural operations to reduce seasonality and provide year-round employment. However, these efforts have been overshadowed by the interest in crop specialization and increased concentration of large farms, encouraged in part by the U.S. Department of Agriculture and the Cooperative Extension System.

In the 1940s DiGiorgio Farms, a large operation in the Central Valley of California, diversified production so that nearly all workers were employed year-round. Most (80 percent) of DiGiorgio's land was in vineyards, 11 percent was in orchards, and 5 percent in vegetables; the remainder was not planted. In 1946 DiGiorgio reported that 92 percent of its workers were permanent employees, 7 percent were temporary, and 0.5 percent were transient (Glover 1984, 274–75).

A 1992 study among thirty-two field crop farmers in California's Central Valley revealed that most made a conscious effort to arrange farm work to provide year-round employment for their regular workers (Vaupel et al. 1995). As part of that study I conducted case studies of four farmers who had successfully created year-round jobs for many of their employees and substantially decreased seasonal employment as a proportion of total employment on their farms. Three of the farmers were in different regions of California, and one was in Pennsylvania (see Table 9.1). These four case studies and the 32 interviews with farmers in California's Central Valley provide the basis for the following discussion of strategies to achieve diversity and year-round agricultural employment systems.

Several trends are now rekindling interest in diversified farming. Diversification is considered a solution to many environmental problems caused by crop specialization. It also is becoming an avenue toward revitalizing agriculture and rural economies (Stauber, Chapter 8) and to capturing premium price opportunities in rural/urban fringe areas (Lapping and Pfeffer, Chapter 7). When planned with labor manage-

Table 9.1. Crops and enterprises on survey farms

Southern Calif. Desert Farm (485 acres)	Central Calif. Farm (7000 acres)	Northern Calif. Farm (225 acres)	Pennsylvania Farm (600 acres)
Crops			
Bell peppers	Alfalfa	Barley	Wheat
Cabbage	Dry beans	Beans	Rye
Herbs	Rice	Oats, hay	Corn
Kale	Safflowers	Wild rice	Oats
Leaf lettuce	Sunflowers	Garlic	Soybeans
Snap peas	Tomatoes, processed	Potatoes	Tomatoes
Sweet corn	Vetch for seed	Tomatoes, fresh	Sweet corn
Squash	Wheat	Raspberries	Green beans
String beans	Tree Fruit	Wine grapes	Cabbage
Tomatoes, fresh		Herbs (11)	Zucchini
			Carrots
Farm enterprises			
Roadside stand	Dried fruit	Bay leaf wreaths	Livestock
Farm labor		Garlic wreaths	Flour mill
contractor		Packaged herbs	Bakery
		Raspberry jam	Cannery
		Sun-dried tomatoes	Freezer processing
		Mail-order sales	Peanut butter roaster
		Electric power	Body care products
		Livestock	Printing/publishing
			Mail order
			Retail store

ment in mind, diversification can lead to year-round employment for many seasonal farm workers.

Advantages of a Year-Round Employment System

Advantages for Farmers

Worker availability. Even with the current surplus of agricultural workers, growers worry about the availability of seasonal labor. Some growers are in areas that are not part of the regular migrant stream. Some have important but minor labor needs when other crops are at their peak. Some worry that when they lay off their trained workers when one task is complete, those workers will be employed elsewhere when they are needed again.

Having year-round workers eliminates these worries. The benefits are even greater during labor shortages, such as might occur from changes in U.S. immigration policy or improvements in the economic situation in Mexico.

Increased productivity. Many of the growers interviewed are convinced that year-round workers are more productive than short-term workers. Year-round workers know the unique characteristics of each field, such as low points that hold water and other areas that receive less water. Long-term workers also have built up skills and experience in each task and an understanding of how the grower wants the work done. They may be more productive because they feel loyal toward employers who offer them year-round, continuous employment.

Most farmers know the importance of having an experienced, loyal, and skilled regular workforce: "A good man makes the work look easy." "[Long-term workers] take better care of the ranch than I do." "Experienced irrigators can make water run uphill."

Less need for training. Growers have found that a steady workforce needs less training. Although year-round workers must learn more skills for diverse tasks, additional baseline training is seldom necessary once these skills are acquired.

Increased personal satisfaction. In many ways, the farmers in the study expressed personal satisfaction from a system that offers workers long-term employment and good wages and benefits. They valued their close relationships with employees. One grower, for example, helped a worker's son apply for a college scholarship to the grower's alma mater. Some growers give employees pay advances and interest-free loans. They are pleased when employees can buy homes or trailers.

Lower unemployment insurance rates. Unemployment insurance is paid to workers who are temporarily out of work through no fault of their own. Rates are determined by the number of claims made by laid-off workers against an employer. In California, rates vary from 0.1 to 5.4 percent of wages. Growers who offer extended or year-round employment instead of laying off workers will lower their unemployment insurance rates and realize savings.

Lower workers' compensation rates. Workers' compensation insurance pays medical expenses for job-related injuries or diseases and partially pays for lost wages in some circumstances. Premiums increase when injured workers make claims against the employer. Studies have demonstrated that the high accident rates in large agricultural firms may be associated with a high degree of worker turnover, which shows that the workers are less experienced. On the other hand, a steady workforce is more familiar with the equipment and farm and therefore less likely to become injured. The farmers interviewed in the case studies had workers' compensation rates well below the standard.

Less damage to equipment. Several growers believe that year-round workers are more familiar with their farm equipment and operate it more carefully, causing less damage than new workers would.

Advantages for Farm Workers

Currently, farm workers bear the costs of the seasonal job structure. Having year-round work means that they no longer must face disruptions from frequent seasonal layoffs, such as job searches, long periods of unemployment, and migrating to look for employment.

Steady work means a higher annual income and can mean more benefits. Workers who earn minimum hourly wages can earn $9,000 to $11,000 a year, well above the $5,000 to $7,500 average for agricultural workers. Most employers offer more benefits, such as health insurance, paid vacations, annual bonuses, and housing, to year-round than to short-term employees.

Year-round work means a higher standard of living for workers and their families. They can keep their children in the same schools and develop ties to the local community. Some workers can buy their own homes or trailers.

Additionally, workers feel that they receive better treatment from long-term employers. All the year-round workers we interviewed were pleased with their jobs, and most said that they would remain as long as the employer wanted them.

Diversity: The Key to Decreasing Seasonality

Among the four farmers I interviewed who had successfully established year-round employment, diversity was the key to solving the seasonality problem. Diversifying production is a strategy farmers often use for agronomic reasons and to help reduce economic risk. The case study farmers found that diversity can reduce the seasonality problem in farm labor by extending the employment period.

We identified three levels of diversity used by agricultural employers with ten- to twelve-month employment systems: diversity of plant varieties, diversity of crops, and diversity of farm enterprises. By planting both early- and late-season varieties of the same crop, a farmer can extend the growing season. This first level of diversity enables growers to extend the production season while using their existing knowledge and skills for growing a particular crop.

Crop diversification, the second level of diversity, allows growers to extend the work season even further. The easiest kind of crop diversification involves adding crops similar to those already grown, such as spinach added to lettuce production, or corn and safflower to wheat production. Greater diversification in the types of crops usually means even longer periods of employment. Growing annual and perennial crops can extend employment to a year-round basis because of the different growing seasons and the off-season tasks required for each.

Soil and climate limit the extent of crop diversity, but more varied crops usually can be grown in a region than are currently being produced. Often, the present crop mix is the result of regional specialization and does not represent the only crops that be grown. For example, Karl Stauber notes in Chapter 8 that Iowa produced more than thirty commercial vegetable crops at the turn of the century.

New technologies can facilitate crop diversification. New varieties have been developed to grow in climates that were previously unsuitable for the crop. Pest-resistant varieties also enable growers to produce certain crops where they could not be grown before. Greenhouses and shades allow crop production in areas thought to be too hot or cold.

Enterprise diversity, the third level of diversity, is the most comprehensive means of extending the employment season. Value-added processing of farm products is a natural extension of farming and a good way to extend the income stream and employment season. Examples include production of garlic braids, jams, gourmet oils, sun-dried tomatoes, cider, dried herbs, dried fruits, and fresh pies.

Additional marketing components can further diversify the farm. A mail-order operation enables the grower to expand sales and receive retail prices while employing workers to fill orders. Roadside stands and selling at farmers' markets provide additional employment and sales.

Nonagricultural enterprises can also be added to farm operations. One farmer in the North Coast region of California employs workers to make bay leaf wreaths in November and December for holiday sales. Experienced wreath-makers earn from twenty to thirty dollars per hour. An Indiana farmer found he could use his well-equipped workshop during the winter months to do machinery repairs for local businesses.

Additional Strategies for Year-Round Employment

Although diversity is key, the labor aspects of every management decision must also be considered in developing year-round employment. Planting new crops, eliminating crops, and changing production technologies all affect labor requirements. Growers with year-round employment avoid peaks and valleys by selecting a combination of crops that requires a constant number of employees throughout the year.

Crops that just break even, or are produced at a loss, are sometimes grown for disease and pest control, but the same concept can also be applied to keep year-round employees. For example, maintaining citrus trees that are harvested in winter enables a farmer to keep workers employed and use machinery that otherwise would be idle. By growing sugar snap peas, a break-even crop, one farmer fills the only gap in employment and keeps his best workers employed continuously for ten months. Sometimes, a crop expected just to break even will turn a profit.

New technologies, such as plastic tunnels, hot caps, and greenhouse starts, enable farmers to grow crops earlier in the season and continue production later in the season. In some regions, farmers can use these technologies to double crop their land.

Although mechanization usually displaces workers, it can also help establish year-round employment when it is done selectively to level out labor requirements. Selective mechanization may mean using less than the most advanced labor-saving technology or continuing to use hand labor to keep workers employed during a particular time of year.

A full analysis of the costs and benefits of mechanization decisions includes not only the usual monetary costs but also the effect on the labor management system. For example, one grower decided that savings from mechanizing his string bean harvest were not worth the bad feelings engendered in his regular workers from losing work. Another farmer bought tomato harvesters but did not add electronic sorters because they would have eliminated the need for many workers and introduced a gap into his continuous employment pattern.

Farmers can use various employment management techniques to increase year-round employment. These include planning and scheduling labor tasks, staggering the planting, pacing the work over a longer time, saving work for the off-season, and coordinating labor needs with other farmers and local industries. Central to each technique is the need to eliminate job specialization and train all workers for multiple tasks.

Growers who are unable to extend employment on their own farms can coordinate with other employers who have work available during gaps in employment. Many farmers share crews. Some refer employees to agricultural processing jobs that are available during the winter. Growers also refer workers to local nonfarm industries. A Farm Bureau publication from 1941 suggested a general industry approach to sharing labor through joint operations and organizing citrus crews to work in other crops such as tomatoes (Smith 1941).

A year-round system has its challenges, but the farmers we interviewed had succeeded in reducing the number of seasonal workers and had found that the effort was worthwhile for the financial health of their farms and for their personal satisfaction.

Beyond Seasonality: Other Aspects of the Vision

Reduced seasonality of employment will not be enough by itself to fulfill my vision for all farm workers. Legal protections must be improved for agricultural workers, and new solutions must be found for many long-standing problems.

Reducing the Rhetoric

Most importantly, we must reduce the rhetoric that separates growers and their advocates from farm workers and their advocates. The shrill pitch of rhetoric in agricultural labor issues leaves little room for resolution. Each side expounds only one version of the picture, yet two realities exist simultaneously.

A giant step toward improving wages and working conditions of farm workers would be for each side to recognize that the other's position has some validity. Employers and their advocates must admit that some growers and contractors exploit farm workers. Only through recognition of this fact, rather than a continuing denial of it, can a cultural norm develop that no longer tolerates such exploitation. On the other hand, farm workers and their advocates must admit that not all growers and contractors exploit farm workers and that some have excellent labor management systems. Growers who provide good wages and working conditions deserve to be recognized and supported.

Having acknowledged that both realities exist, the two sides can begin a dialogue on the real issues. Growers, grower organizations, workers, and worker advocates can come together to condemn poor treatment and exploitation of farm workers and make it clear that such practices are not acceptable. They can also identify and promote exemplary labor management systems. These explicit statements can greatly affect the lives of farm workers.

Rethinking Labor Protection Laws

Federal and state labor laws provide workers with many protections, such as minimum and overtime wages, field sanitation, equipment safety standards, pesticide safety standards, and in some states, the right to elect a representative to bargain on their behalf. Current laws are far from perfect, but the major shortcoming is that they are not being enforced. Neither federal nor state agencies have adequate resources for enforcement, nor are they likely to receive such resources soon. However, the current enforcement system can be made substantially more effective, even at current funding levels.

First, state laws must be brought into conformity with federal laws in areas where they overlap. Often, slight differences in definitions or details keep the two enforcement systems apart. If the two are reconciled, enforcement efforts can be combined and resources will be increased accordingly. State laws should continue to be allowed to provide greater worker protections than federal laws.

Second, available resources must be concentrated in a serious, systematic, and visible enforcement effort in the field. Intensive sweeps by combined federal and state agencies in California have encouraged many growers and FLCs to respect labor protection laws for the first time. Rather than conducting the usual random investigations, enforcement officers have focused on one region and systematically visited growers and contractors at their business addresses. This joint, concentrated enforcement has increased the likelihood that violators will be fined and has resulted in improved efforts by agricultural employers to avoid violations. It has also benefited growers who follow the law and have been at a competitive disadvantage compared with those who do not.

Third, education about labor laws must be improved to make enforcement more effective. Some violations are the result of violators' lack of knowledge about and understanding of the laws. Although some grower organizations disseminate information about labor laws, many farmers and most labor intermediaries do not receive this information. Enforcement agencies must find appropriate channels to educate farmers and FLCs in the languages that they speak.

Fourth, all labor protection laws should be expanded to cover farm workers. Most federal labor standards and protective laws originally exempted agriculture. Exemptions have since been eliminated for minimum wages, social security, and unemployment insurance, but in most states eligibility standards still exclude many agricultural workers from unemployment insurance. Only a few states have extended unemployment insurance and workers' compensation insurance to cover most agricultural workers.

Agricultural operations are also exempted from the National Labor Relations Act, which gives workers the right to elect a representative and bargain collectively with employers. In 1975 California passed the Agricultural Labor Relations Act, which includes special provisions adapting the national labor relations framework to the unique aspects of agriculture. Several other states have adopted labor relations laws for agricultural workers, although none is as comprehensive as California's.

Historically, unions have played an important role when wages and working conditions decreased to intolerable levels. In my vision for fair treatment of agricultural workers, unions will be an accepted part of the agricultural labor system. Growers will have learned to live with unions and unions will have learned to bargain responsibly and effectively with growers. Arturo Rodriguez, president of the United Farm Workers, has expressed this new attitude, claiming that the union will "do everything possible to ensure that unionized growers flourish" (*California Farmer* 1996).

Continue and Expand Services

Many farm workers do not have adequate housing, health care, transportation, schooling for their children, child care, English language training, or training in special skills. These problems are greatest for migratory workers and their families. If farm workers received adequate living wages and benefits, special services would not be necessary. Until these conditions are achieved, it is important to continue and expand services for farm workers and their families.

The major farm worker assistance programs are Migrant Education, Job Training, Migrant Health, and Migrant Head Start. Overall, the scale of these programs is inadequate to address the scope of the problems. For example, the Department of Labor reports that it serves about thirty thousand farm workers annually in its training programs, less than 1 percent of the total. In fiscal year 1996, the USDA's Rural Housing Service received funding to build only 550 units nationwide (Martin et al. 1996, 10), or one unit for every seven thousand workers.

As with labor law enforcement, budget cuts leave little hope for expanding governmental services. Instead, we must rethink each problem. Local solutions will be required, since federal and state services are often not available. Private enterprise must be used as a vehicle for delivering many services. Involvement of entire communities, including farmer-employers, will be necessary.

Following are a few examples of new ideas that have been tried and others that are possible but have not yet been tested.

Housing. The number of housing units for farm workers has decreased over the past thirty years for many reasons. Enforcement of local health codes has added costs for upkeep and renovation and has resulted in the closure of many units. Few new units are being built because of funding cuts and the reluctance of farmers to make the necessary investment. Funding has also been reduced for low-cost housing in general, including units that settled farm workers could rent or buy.

New ideas for farm worker housing are being developed and implemented at the local level. In California, one company has designed house trailers to accommodate single male workers or farm worker families. Straw bale construction covered with adobe offers another low-cost alternative and has recently been approved for housing in California. At least one community has developed a new farm worker housing project with community services built into the plan.

Child care. Women make up about 29 percent of the agricultural workforce (Mines et al. 1991, 40). When they have no child care, women must turn down employment opportunities or bring their children to the fields. With start-up assistance, small child-care businesses or cooperatives for the children of farm workers could be started by women farm workers, former farm workers, farm worker wives, or other family members who are not currently in the workforce. Child-care sites should be operated at convenient locations for the workers, such as at the work site, at farm worker housing projects, or in neighborhoods where farm workers live.

English language training. A survey in Ventura County on the southern coast of California found that farm workers, especially women, feel their options are limited because they do not speak English. Almost half rated lack of English as the primary barrier to finding nonagricultural work, and some see it as a barrier to finding more desirable jobs in agriculture (Vaupel 1992). Yet few English classes are available and convenient for farm workers. Transportation to classes and finding child care are additional problems. English classes would be better attended if brought to the workers. Classes could be held at the work site, within housing areas or neighborhoods where farm workers live, at churches, or together with other training classes.

Community Changes

Besides the problems I have addressed at the farm level and in the legal setting, farm workers have problems at the community level because of their low incomes and the discrimination they experience as immigrants. Many people in the community perceive immigrants as the cause of widespread economic and social problems. Unsubstantiated reports claim that illegal immigrants rob American workers of highly paid jobs and that they claim billions of dollars of benefits from the government.

This perception differs significantly from the results of most economic studies, which consistently find that the overwhelming majority of undocumented Mexican workers are in the lowest skilled, lowest paid occupations, such as agriculture, the garment industry, and the hospitality industry. Undocumented workers also pay more in taxes and social security deductions than they receive in public services. Studies have found that fewer than 4 percent of illegal immigrants had children in U.S. schools. Only 0.01 percent to 0.4 percent of welfare recipients were undocumented immigrants, and many of these were legally eligible for benefits (Bustamente 1981, 99–100).

For more than a century, the farm labor market has been a last resort that most Americans try to avoid. Field work offers lower wages and much harder manual work than most nonimmigrants will tolerate. Farmers will quickly assert that workers born in the United States cannot work as hard as Mexican-born workers.

To correct the general perception of immigrants, it is necessary to begin a public dialogue on the objective facts and to recognize the contributions that immigrants make to the economy, especially the important role they play in food production. Farm workers must be accepted as part of the communities in which they live. This is already being done in some rural areas, where local people, including farmers, have expressed a sense of community with farm workers and have advocated for more services for them.

Conclusion

U.S. agriculture depends on immigrant workers. Despite marginal improvements over the past fifty years, the living standards of most farm workers are far below those of nonagricultural workers. Some farm workers are victims of outright exploitation. The benefits of past strategies—labor protection laws and government services—have probably reached a plateau. Meanwhile, farm workers' real wages and benefits are declining.

To address the root problem adequately, we must focus on the very characteristic that distinguishes farm work from industrial work: seasonality. The need for a seasonal labor force in agriculture developed with specialized production and large concentrated farms. Seasonality brought the need for a fluid workforce that would disappear between tasks. Seasonal workers must piece together a series of short-term jobs, but most are unable to earn a living wage during the average twenty-nine weeks of agricultural work available to them each year.

By restoring diversity into farming systems, both large and small farms can offer year-round employment that will enable farm workers to make a decent annual income and settle into the community with their families. Large and small farms alike can plant a variety of crops and add diverse enterprises to their farm operations. By developing year-round employment systems, farmers no longer will need to worry about worker availability and will find that full-time workers are more productive than short-term workers.

To fulfill this vision, changes will also be made in the legal setting and at the community level. Protective labor laws will be expanded to cover all farm workers, and enforcement will be more effective because state and federal laws will be consistent and enforcement agencies will combine their resources in concentrated enforcement efforts. Farm worker services will be maintained and expanded to provide necessities that workers are unable to buy with their low seasonal wages. Public-private partnerships will develop to fill the gaps in services and expand their effectiveness. Finally, a public dialogue will develop to defend the contributions of immigrants to the U.S. economy and culture. Communities will accept and respect the farm workers that live among them.

To accomplish real changes that improve the lives of farm workers, we as a country will alter our expectations, no longer accepting substandard living conditions for those who labor to produce our food. In 1967, Varden Fuller observed, "We share with other industrialized countries the evolution of the philosophy—with some backing in explicit policy—that farmers should have incomes equal to those that prevail in comparable occupational categories, but Americans have not gone that far for hired farm workers." (98)

My vision extends this philosophy to farm workers so that they will have the incomes, benefits, living conditions, fair treatment, and respect that prevail in comparable nonagricultural occupations. This is a modest but realistic vision that will make significant improvements in the lives of millions of farm workers and their families.

Reference List

Bustamente, Jorge A. 1981. "The Immigrant Worker: A Social Problem or Human Resource." In *Mexican Immigrant Workers in the U.S.*, edited by Antonio Rios-Bustamente. Los Angeles: Chicano Studies Research Center Publications.

California Farmer. 1996. UFW lands the biggest vegetable contract. 279 (8): 47.

Fritsch, Conrad F. 1984. "Seasonality of Farm Labor Use Pattern in the United States." In *Seasonal Agricultural Labor Markets in the United States,* edited by Robert D. Emerson, 64–95. Ames: Iowa State University Press.

Fuller, Varden. 1940. "The Supply of Agricultural Labor as a Factor in the Evolution of Farm Organization in California." In *Hearings before a Subcommittee* of the Committee on Education and Labor, 19777–898. U.S. Senate, 76th Cong., 3d sess., 1940.

———. 1967. "Farm Manpower Policy." In *Farm Labor in the United States,* edited by C. E. Bishop, 97–114. New York: Columbia University Press.

———. 1991. *Hired Hands in California's Farm Fields.* Giannini Foundation Special Report. Davis: Division of Agriculture and Natural Resources, California Agricultural Experiment Station.

Glover, Robert W. 1984. "Unstructured Labor Markets and Alternative Labor Market Forms." In *Seasonal Agricultural Labor Markets in the United States,* edited by Robert D. Emerson, 254–79. Ames: Iowa State University Press.

Mamer, John W. 1984. "Occupational Structure and the Industrialization of Agriculture." In *Seasonal Agricultural Labor Markets in the United States,* edited by Robert D. Emerson, 286–323. Ames: Iowa State University Press.

Martin, Philip, J. Edward Taylor, and Michael Fix. 1996. *Rural Migration News* 2 (April).

Mines, Richard, Susan Gabbard, and Betsy Boccalandro. 1991. *Findings from the National Agricultural Workers Survey (NAWS) 1990: A Demographic and Employment Profile of Perishable Crop Farm Workers.* Washington, D.C.: U.S. Department of Labor.

Mines, Richard, Susan Gabbard, and Ruth Samardick. 1993. *U.S. Farmworkers in the Post-IRCA Period.* Washington, D.C.: U.S. Department of Labor.

Rosenberg, Howard, Susan Gabbard, Eric Alderete, and Richard Mines. 1993. *California Findings from the National Agricultural Workers Survey: A Demographic and Employment Profile of Perishable Crop Farm Workers.* Washington, D.C.: U.S. Department of Labor.

Smith, Roy J. 1941. Labor stabilization in the citrus industry. *Los Angeles County Farm Bureau Monthly* XV (7): 16D, 30.

U.S. Bureau of the Census. 1994. *1992 Census of Agriculture.* Vol. 1, Part 51: United States Summary and State Data. Washington, D.C.: U.S. Department of Commerce.

Vaupel, Suzanne. 1992. *A Study of Women Agricultural Workers in Ventura County, California.* Ventura: Committee on Women in Agriculture.

Vaupel, Suzanne, and Philip L. Martin. 1986. *Activity and Regulation of Farm Labor Contractors.* UC Giannini Series No. 86-3, Davis: Division of Agricultural and Natural Resources, University of California.

Vaupel, Suzanne, Gary Johnston, Franz Kegel, Melissa Cadet, and Gregory Billikopf. 1995. *How to Stabilize Your Farm Work Force and Increase Profits, Productivity, and Personal Satisfaction.* Davis: University of California Sustainable Agriculture Research and Education Program.

10

Building Coalitions for Agriculture, Nutrition, and the Food Needs of the Poor

Beatrice L. Rogers

More than 14 million children under the age of eighteen were living in poverty in 1994; an estimated 30 million Americans experience hunger because of resource constraints at some time during the year; an estimated four million children under twelve in low-income households go without sufficient food at least sometime during the year; one-fifth of low-income households with children are reporting hunger at least part of the year because of financial constraints (Center on Hunger, Poverty, and Nutrition Policy 1992; Food Research and Action Center 1995; May and Porter 1996, 22). The statistics showing the continued need for food assistance for the poor in the United States are as unarguable as they are appalling. A coalition of farm- and food-industry interests, nutrition advocates, and antipoverty, antihunger forces has built a network of programs providing food aid to the poor; this coalition could strengthen them, educating the public to recognize that these programs are a "win-win-win" proposition, contributing to the farm and food economy, alleviating hunger, and building human capital through improved educational productivity, nutritional status, and health.

The current mix of farm- and food-assistance programs does not represent a coherent national food policy. Programs have emerged as a series of responses to current conditions, modified through political compromise to obtain support; they have developed their own interest groups that help to preserve them, trading support with other interest groups when necessary. The existing coalition between farm interests and antihunger advocates has been an uneasy one (Clancy 1993), with generally conservative farmers often expressing opposition to welfare

139

benefits for the poor, even though legislators and lobbyists recognize the necessary links between them. Nonetheless, there has been a shift in the balance of funding within the U.S. Department of Agriculture from programs that are intended to support agriculture but have a collateral benefit to the poor toward programs specifically designed to aid the poor but that also benefit the food sector of the economy (Clancy 1993).

Federal food-assistance programs have made a great difference for the food security of the poor. The largest of these, the Food Stamp Program and WIC (Special Supplemental Program for Women, Infants and Children), now reach most people eligible for their services. But even these programs leave some eligible individuals unserved, and in many food-assistance programs, including school meals, coverage is incomplete, outreach inadequate, and funding insufficient to cover all eligible recipients. The programs are politically vulnerable, and the level of support is constantly at risk of being reduced. In 1996, some emergency programs saw their budgets cut; even the entitlement status of Food Stamps was threatened, as both Food Stamps and WIC for a while appeared subject to major reductions.

These programs were saved from what would inevitably have meant severe cutbacks in coverage and level of benefits by the actions not only of farm lobbyists and farm state legislators (who historically have been advocates for programs that expand the demand for food) but also of food-industry lobbyists, who for the first time added their influence to that of the nutrition and antipoverty advocacy groups to protect the Food Stamp Program. This coalition of interest groups clearly showed it has the power to protect food-assistance programs.

Thus the vision I offer entails defending and expanding food-assistance programs by building a coalition based on an explicit recognition of the multiple constituencies the programs serve. This requires marshaling evidence not only of the nutritional effectiveness of each program but also of its effect on local and national farm and food economies. It requires lobbying in its most positive sense: educating the public, legislators, and the food industry so that they join with farm lobbyists and antihunger groups in supporting these programs because of the wide-ranging benefits they provide.

The number of farms in the United States is shrinking, but the food sector of the economy still accounts for one in six jobs and about 16 percent of the gross domestic product (Community Nutrition Institute 1995a). Food assistance increasingly is provided as added purchasing power for retail food rather than by direct distribution of farm commodities, so not only farms, but the entire food industry benefits from these programs. The Food Stamp Program, for example, expanded demand for food by as much as $4.2 billion in FY 1988 (Senauer et al. 1991), of which only about 21 percent was for farm commodities (Elitzak 1993). This represents about 1 percent of total demand in that year (U.S. Bureau of the Census 1995, Table 1099). Other estimates of the effect of Food Stamps on food demand range from 1 to 2.6 percent (Clancy 1993).

Linkage of Farm Support and Food Assistance to the Poor

Farm price supports and food-assistance programs have been linked since the origin of both during the Depression. Agricultural surpluses coexisted with widespread hunger, and there were several dramatic instances of food being destroyed because it could not be sold. The perception of food going to waste, even being destroyed, while millions of Americans went hungry gave political impetus to the development of programs to provide food to the poor both through nonmarket programs, such as surplus food distribution, and through programs such as Food Stamps, which increased market demand for food. It is not a coincidence that the Depression fostered not only economic support for farmers but also social welfare programs designed to provide food and income support to poor families. The farm economy was devastated by the loss of consumers' purchasing power at a time when a quarter of the labor force was unemployed. The logic of helping both farmers and poor consumers through food assistance was compelling.

Total agricultural production in the United States was neither lower nor higher than average during the Depression years. The widespread surpluses seen as evidence of oversupply arose from the decrease in effective demand among both American and overseas consumers. Still, the problem of surpluses was widely perceived to be remediable through supply controls combined with price supports. For the next six decades, it was common for the federal government to maintain floor prices for selected agricultural commodities by committing to purchase surplus stocks whenever necessary to raise prices to the support level. This commitment often put large stocks of certain commodities in government hands. Over the years this has led to the perception that agricultural surpluses were a natural part of the American agricultural system and that these surpluses should be put to use in humanitarian ways to assure the adequate nutrition of poor people.

The relative importance of agricultural interests as a driving force behind food programs has diminished as the antipoverty, antihunger movement has become stronger. Perceptions of when it is appropriate for government to intervene in the market have changed: in contrast to the Depression era, today alleviation of poverty is accepted as a legitimate public goal. This change is paralleled in the changing view of the major purpose of food assistance, from surplus disposal to demand expansion as a way of increasing individual economic well-being and welfare.

In recent decades, the emphasis has shifted from assuring consumption of an adequate quantity of food toward promoting consumption of particularly nutritious foods. Food has always been seen as a "merit good," more worthy of public subsidy than other kinds of consumer goods because it meets the most basic of human needs. But food-assistance programs now are being designed according to a more "medical model" of the role of food, and of specific foods, in promoting good health and preventing disease.

Possibly the newest phase in the evolution of food-assistance programs is toward the provision of emergency aid through soup kitchens and emergency food pantries (Poppendeick 1995). Federal food is being channeled to these primarily private charitable outlets in response to a perception that the hunger related to severe poverty and homelessness is a temporary emergency. This is a dangerous development. Household food security depends not on the temporary availability of charity but on a reliable source of income to buy food.

As Poppendeick (1995) points out, new objectives do not displace old ones as food programs evolve. Rather, multiple objectives coexist and, it is hoped, expand the range of groups with an interest in preserving the programs.

The Shifting Emphasis in Food-Assistance Programs

Two major food-assistance programs were started during the Depression: Surplus Food Distribution, which distributed to poor families the food purchased by the Commodity Credit Corporation to support prices, and a Food Stamp Program, which was quite different from the one we know today. The original Food Stamp Program, like Surplus Food Distribution, was designed to increase demand for specific foods whose prices fell below the mandated support price. To ensure that the food stamps increased the total demand for food, there was a purchase requirement: families paid a certain amount of money and received stamps worth that amount that were restricted in use to food; they also received "bonus" stamps, distinguished by color, which could be used only for officially declared surplus foods. Both of these programs benefited the needy, but their design emphasized surplus disposal by restricting the range of commodities obtainable.

The original Food Stamp Program ended in 1943. In the booming wartime economy there was no justification for government-supported demand expansion, and jobs at good pay were plentiful. But the Surplus Food Distribution program continued through the early 1970s (it still operates on Indian reservations), and several other programs still provide surplus foods intermittently through various poverty programs. In 1946, the National School Lunch Program (NSLP) was initiated, partly in response to the discovery that up to one-third of military recruits were rejected for health reasons related to inadequate nutrition. Thus, it was partly a response to nutritional concerns, but again, the food came largely from surplus stocks.

The NSLP's direct link to surplus disposal has not always benefited its nutritional objectives. For example, surplus commodities provided to the program include canned hamburger, peanut butter, cheese, and butter, contributing to the justified criticism that school meals are excessively high in fat and sodium and deficient in fresh fruits and vegetables. Similarly, milk is always provided (because of strong support for the dairy industry) and is consistently identified as a major nutritional benefit of these programs (Hanes et al. 1984; Devaney and Fraker 1989; Gordon and McKinney 1995). Only recently, however, has it been permissible to serve low-fat or skim milk in the school meals programs, despite the association of whole milk with high fat intake.

The NSLP nicely demonstrates the shift in emphasis from surplus disposal to poverty alleviation and then to nutritional objectives. In the 1960s and 1970s, attention was focused on systems to provide free and low-priced lunches to low-income children without the stigma of being identified as poor. In 1966, the School Breakfast Program was initiated primarily to serve needy children. In the past fifteen years, guidelines for school feeding programs have focused on the provision of healthier meals. Starting in the late 1970s, the USDA began to experiment with offering "cash in lieu of commodities," giving flexibility to school food-service personnel to choose more nutritionally desirable foods. In 1995, new NSLP guidelines were legislated that required the meals to conform to the U.S. Dietary Guidelines, notably by providing more fruits and vegetables.

The major food-assistance program in the United States is the current Food Stamp Program. This program emerged in the 1960s and saw its greatest expansion and its establishment as an entitlement during the height of the War on Poverty. The Food Stamp Program provides cashlike benefits to poor households; these benefits can be used for any food, irrespective of its nutritional value. It acts much more like an income-support program than one for surplus disposal, although food stamps still increase food purchases two to four times more than the same value in cash (Senauer and Young 1986; Fraker 1990). When the program started in 1964, expansion of food demand was more significant, with the program requiring participants to pay for their stamps. As in the original Depression-era program, participants were required to "lock" some of their own money into food purchases, and in return they received bonus stamps, worth more than their own cash contribution. A critical difference was that in the modern program, any food, not just foods in surplus, could be bought with the stamps. In 1979, the requirement to purchase stamps was eliminated. This change has made it even more like an income-support program and less like one serving either surplus disposal or specific nutritional goals.

In fact, the Food Stamp Program meets many criteria for a well-designed income-support program: benefits are provided to all the poor, irrespective of their other characteristics, such as age, sex, family status, or residence; benefits are on a sliding scale varying inversely with income; and the level of benefit is closely tied to a cost-of-living indicator, the Thrifty Food Plan, so that the income transfer is adjusted for inflation.

Politically and practically, an important difference between the Food Stamp Program and Surplus Food Distribution is that the former uses the entire food marketing system. As a result, processors and marketers benefit from the program even more than the farm sector. Where Surplus Food Distribution bypassed the food market, the Food Stamp Program strengthens it, creating a new interest group helped by the program. The importance of this characteristic was demonstrated in 1996, when the Food Stamp Program was threatened with cuts through block granting.

The Food Stamp Program has been criticized because it makes no effort to alter food choices in nutritionally desirable directions and recipients can buy any foods they choose. Compared with other poor consumers, however, Food Stamp recipients choose foods that provide more nutrients per dollar spent (Fraker 1990). (Poor consumers in the United States obtain more nutrients per dollar spent than does the general population; Food Stamp users do better still.) Still, the freedom of choice given to Food Stamp users has been seen as nutritionally disadvantageous, since families may make unwise choices. Polls often find public support for the Food Stamp Program lower than support for the more nutritionally focused WIC program, for example, presumably because of this perception. This is ironic, since the association of Food Stamps with food as a merit good has been a source of its political strength. For example, when the Nixon Administration's proposed

Family Assistance Plan was defeated because benefits were excessive, Food Stamp Program benefits were raised. The Food Stamp Program provides a clear demonstration of the strength that comes from providing income support, nutritional benefits, and demand expansion for both farm commodities and retail foods.

As the Food Stamp Program was a natural outcome of the focus on poverty alleviation in the sixties, so the WIC program, initiated in 1972, reflects an emphasis on the health and nutritional benefits of food assistance. WIC provides generous food supplements to pregnant and lactating women and infants and children from low-income households. The supplementary food represents a substantial contribution in income terms, but the range of foods is strictly limited to specific items chosen to help meet the nutritional demands of pregnancy, lactation, and the rapid physiological development of childhood. The program follows a medical model, with supplements provided in the context of medical care and nutrition education. It enjoys strong public support because it is unambiguously associated with health and nutrition, the supplementary foods all are nutritionally recommended, and the beneficiaries are women and children. Yet WIC, too, adds significantly to food demand. Infant formula companies have competed strongly to receive the contract for providing formula to the program. Cereal companies have lobbied for more of their products to be accepted. (Currently, cereals must be iron fortified and limited in sugar to qualify for purchase with WIC coupons.) Clearly, the program represents both a health intervention and big business for the private sector.

A recent innovation in the WIC program has been the Farmers Market Nutrition Program, which provides ten dollars worth of coupons each season for use in farmers' markets. Although the amount of money is quite modest, the effect on individual farmers in particular local areas has exceeded the value of the benefit provided (Community Nutrition Institute 1995b). This program no doubt owes its adoption to the fact that it is associated with both the promotion of local agriculture and the provision of nutritionally desirable fresh fruits and vegetables. As with the larger programs, it shows the effectiveness of serving multiple constituencies.

Clancy (1993) has noted that the sustainable agriculture movement has been absent from the coalition to promote food assistance to the poor. This situation may be changing. In 1996, the Community Food Security and Empowerment Act was passed, with funding of $1 million for the first year, to promote the food security of low-income groups through community gardens, direct farm-to-consumer links, and other community-based efforts. Groups advocating for local, small-scale, and sustainable agriculture were active in the promotion of this legislation. The program is small and cannot serve as the same kind of guarantee of family food security as an adequately funded and secure Food Stamp entitlement. Still, its success is remarkable in a time of threatened cuts in both farm- and food-assistance programs.

The Farm Programs

The price-support programs begun in the Depression were targeted to only a few commodities, chosen on the basis of their economic importance rather than their nutritional value: wheat, rice, feed grains, cotton, tobacco, peanuts, sugar, and dairy products. The crop price-support programs functioned primarily by "purchase and diversion"; that is, the government would purchase at the support price whatever production could not be sold at that price on the open market. These commodities then needed to be diverted to noncommercial channels.

Agricultural price supports have come under increasing criticism from outside agriculture and from within the sector. The USDA report *A Time to Choose* (U.S. Department of Agriculture 1981), written at the behest of then-Secretary of Agriculture Robert Bergland, was path-breaking in its analysis of the inequitable distribution of price-support benefits, the ineffectiveness of supply controls, their distorting effects on agricultural production decisions, and their role in reducing American agriculture's competitiveness in world markets. Most notable was that this report came from within the Department of Agriculture, historically a defender not only of price supports but also of the existing system for providing them.

Farm price supports have been both evolving in form and decreasing in size for many years. A more market-oriented system of target prices with direct payments has gradually replaced the system of having the government acquire surpluses. The level of the target price has also moved downward, so that market prices often exceed the target price. The 1985 Farm Bill limited how much money any individual farmer could receive (although farmers with more than one farming operation could receive multiple payments, effectively raising the ceiling).

In economic terms, the rationale for eliminating the gross distortions in the agricultural sector, concurrently reducing government-held surplus stocks and cutting costs, is unexceptionable. Only certain crops were supported, and the programs discouraged farmers from responding to market changes. Supply controls as implemented by the USDA have been ineffective in restricting production and even less effective in holding prices high because of the inherent incentive to intensify production when land in cultivation is limited (Paarlberg 1980). High guaranteed prices also reduced the competitiveness of U.S. products in the world market and provided an "umbrella" price that protected foreign producers and encouraged substitution. The high support price for cotton has been an economic advantage for the development of polyester, and the high support price for sugar increased the use of high-fructose corn syrup instead. Clearly, the long-run rationale for commodity price-support programs is shaky. Yet these programs have created an interest group that stands to lose if price supports are eliminated.

Historically, farm programs have benefited from the perception that just as food is a merit good in consumption, so, too, food production is a "merit activity," worthy of public-sector support because farming was perceived as a particularly worthy way of life. However, agriculture in the United States is in possibly the last stage of a process of "losing its uniqueness" as a sector of the economy (Paarlberg 1980). Farmers now make up less than 2 percent of the labor force, and their lifestyles are not much different from those of the nonfarm population. The idealized concept of farming as an activity of self-employed, independent, self-sufficient family units is giving way to a wide array of commercial arrangements including contract farming and farmers working as salaried managers for landowning corporations. Furthermore, most farmers and farm household members engage in paid, off-farm economic activities (Brooks and Reimund 1989). It is increasingly difficult to justify the idea that farming is uniquely worthy of public support. Thus the gradual erosion of farm price supports and their attendant surpluses was perhaps inevitable.

The process of change, whether evolution or radical reform, creates both risk and opportunity. As the surpluses that created the perception of food as a free good diminish, there is a risk that support for programs providing food to the poor also will erode. There is a counter-argument, however. As the agricultural sector ceases to be the beneficiary of price supports, there may be greater incentive for farmers to support demand-expansion programs as a legitimate area of government spending. In the current political climate, cutbacks in all the programs are threatened, and the elimination of entitlements, even Food Stamps, is a real possibility. In this environment, it is more important than ever for farm interests to unite with consumer and antipoverty advocates. Poor consumers need economic support to ensure that they can obtain adequate and acceptable diets; the farm and food sector needs expanded demand in the market.

The Federal Agriculture Improvement and Reform Act of 1996 has been hailed by some as revolutionary in eliminating most of the commodity price-support programs that have been in place in some form since the Depression. But it leaves some large programs almost untouched (dairy price supports remain in place, and sugar and peanut price supports were reduced only moderately). Furthermore, it continues to subsidize farmers, according to a formula based on historical production rather than economic need, by making generous "transition payments" to those who previously benefited from price-support programs. Payments are planned to continue until 2002, totaling $35.6 billion.

These funds currently are planned simply as direct payments to farmers, presumably to compensate for their loss of income from price supports or deficiency payments. These payments, however, will be maintained regardless of any change in world market prices for the commodities. Since world prices for wheat, rice, corn, and other feed grains are currently at historically high levels, these transition payments could exceed the deficiency payments they have replaced, which would have fallen because of high market prices.

Furthermore, even for commodities that have been "decoupled," farming is not fully free. Although the bill that eliminated price-support payments was first called the "Freedom to Farm" Bill, beneficiaries of transition payments are not permitted to grow fruits and vegetables on the land released from supported commodities. This provision is the result of intense lobbying by fruit and vegetable growers' groups who argued that other farmers in effect would be subsidized to compete with current fruit and vegetable producers. But since the payments are received no matter what is produced (or even if nothing is produced) on the land, the bill contains no incentives beyond those of the market to increase production of fruits and vegetables in preference to anything else. Nonetheless, the provision stands.

A Proposed Program

As an example of how antihunger interests might converge with those of farmers and the food industry, I offer a proposal that might provide a politically feasible way to increase the level of benefits in existing food-assistance programs. There are three elements to the concept: supporting American agriculture in its transition from direct subsidies for particular crops to a more market-oriented system; increasing low-income consumers' access to healthy food; and promoting consumption of foods that provide health benefits to consumers and high free-market profits to producers. The proposal is to develop a coordinated program that would encourage the consumption of fruits and vegetables through added purchasing power, possibly through "nutrition stamps," to be distributed through existing food-assistance programs.

Most Americans do not consume sufficient quantities of these foods, which have significant health benefits. Consumption of fruits and vegetables is lowest in the low-income population (Federation of American Societies for Experimental Biology 1995). Demand expansion would focus on low-income consumers, both because they are in need and because their demand responds more to changes in income than does that of the higher income population. In the lowest income quintile, about 42 percent of households' disposable income is devoted to food (Manchester 1991), and a significant portion of any increase in income is likely to go to increasing the total value of food consumed (Fraker 1990). Also, demand for fruits and vegetables responds more to income or price change than does the demand for staple foods. Thus, from the viewpoints of both consumer welfare and the food economy, there is good rationale for targeting demand-expansion programs to low-income consumers.

Fruits and vegetables are not dietary staples. They provide dietary variety and essential micronutrients, but they are not a major source of calories. Households so poor that they cannot consistently afford enough food to avoid hunger would not choose fruits and vegetables as their highest priority purchase, nor should they, because these foods are not calorie-dense. The program proposed here would make a contribution to such households by increasing their purchasing power. The "nutrition stamps" would free them to use other income (such as Food Stamps) for staple foods, while improving the nutritional quality of their diets through the addition of micronutrient-rich foods.

In the aggregate, consumption of fruits and vegetables in the United States increases only a little when the price falls (Finke and Tweeten 1995), and the same is probably true when purchasing power increases (Alderman 1986). In addition, factors other than purchasing power—such as time constraints, lack of knowledge about the preparation of these foods, increasing dependence on food consumed away from home, and consumer tastes and preferences—currently limit consumption of fruits and vegetables.

Nevertheless, low-income consumers are generally far more responsive to both price and income changes than are the relatively better-off. This means that a program of demand expansion targeted to the poor would be more effective than a general price decrease, especially if complemented by promotional and educational interventions.

The proposal can be legitimately criticized as excessively paternalistic, unduly directing the consumption choices of poor people when what makes the most sense is to provide the poor with an adequate income and allow them to make their own consumption decisions. This would provide true food security to poor households. In the current political climate, however, cash transfers such as Aid for Dependent Children are being reduced and restricted; Food Stamps and WIC have survived precisely because of their links to food and nutrition and the fact that they serve the interests of the food industry as well as of the poor. A guaranteed adequate income for all should be the goal of a just society, but meanwhile, we should seek ways to channel purchasing power to the poor in politically feasible ways. Opponents of the federal nutrition programs have used their nutritional shortcomings as a justification for opposing the programs. The proposed intervention, focused on nutritionally desirable foods, would be defensible on both nutritional and welfare grounds.

The structures for distributing such a benefit are already in place. For example, they could be provided as an add-on to Food Stamp benefits and as an addition to the coupons provided through WIC (similar to the Farmers Market Nutrition Program). Although the Food Stamp Program has no provision for nutritional education and behavioral change, the WIC program is well-suited to providing education along with the coupons, linking the nutritional importance of fresh foods with ways to prepare and serve them. Besides serving individuals and households, a similar program could be offered to feeding programs such as schools, emergency pantries, and soup kitchens for the homeless. Providing additional purchasing power to schools would make it easier for meal planners to comply with new regulations to bring school meals closer to the guidelines. In contrast to programs influencing households' private behavior, benefits to schools can easily be tied to specific requirements regarding meal composition, especially in light of recent legislation.

This program of demand expansion would be complemented by a program of technical and financial assistance to farmers to help them switch from program crops to the production of high-value fruits and vegetables, in regions where such production is technically and economically feasible and ecologically sustainable. Promotion of fruit and vegetable production would be aimed at farms currently producing program commodities. The rationale for a regional focus to the agricultural component is that some regions are already producing fruits and vegetables with no assistance; other regions clearly will (and should) continue to produce program crops even without price supports. Therefore, the program will be targeted toward those regions where it is appropriate to shift from program crops to fruits and vegetables.

The United States currently imports 38 percent of the fruits and 7 percent of the vegetables it consumes (U.S. Bureau of the Census 1995, Table 1120). This is clearly linked to seasonal and climatic differences more than to overall production limitations in the country, because we export slightly more fruits and vegetables than we import (U.S. Bureau of the Census 1995, Table 1328). Any increase in demand for fruits and vegetables in the United States could be met by either an increase in U.S. production or an increase in imports, depending on market conditions. In the current era of free trade, under the terms of the North American Free Trade Agreement (NAFTA) and the General Agreement on Tariffs and Trade (GATT), it is unlikely that any program of demand expansion could legally be restricted to American-grown foods. U.S. growers would benefit only in proportion to their share of the fruit and vegetable market. However, there is clearly potential for increased production of fruits and vegetables in the United States. Furthermore, fruit and vegetable production provides the greatest potential to increase the value of farm production (Finke et al. 1995). It could create multiplier effects in the economy because of the need for new storage, transportation, and distribution systems as well.

The idea of increasing consumer demand through income transfer as a way to replace agricultural support programs is not new. In 1968, a USDA analysis concluded that an unreasonably high amount of money would be needed as income transfers to replace fully the income transfer to agriculture achieved through price supports and supply controls (Egbert and Hiemstra 1969). However, the present proposal is not intended to replace price supports but to make use of the resources allocated to the transition in a way that would serve multiple objectives.

Funds made available from the reduction or elimination of price supports might have been captured for the nutritional benefit of low-income households. Moreover, a focus on fruits and vegetables would have the combined attraction of addressing a known nutritional need of the U.S. population, one that is demonstrably greater among the poor, while promoting consumption of foods that are highly profitable for farmers. The proposed program will not resolve the nation's hunger problems nor guarantee the economic health of the farm sector, but it could contribute modestly to both.

The proposal may have become more visionary (that is, less realistic) with passage of the 1996 Farm Bill, however. Allocating transition payments according to historical benefits has created an interest group that would oppose diversion of these funds to other uses. The opportunity to take advantage of these funds has been lost for the next seven years, as the transition payments in their current form represent a commitment to farmers who signed up for them. And the prohibition of fruit and vegetable production is a politically difficult obstacle to overcome.

Transition funds, in their present form, serve primarily to blunt opposition to the gradual elimination of price supports. If they allow us to rationalize and modernize our system of agriculture, they may be worth it. But they could be used more productively in a program that combined agricultural support with demand expansion. Such a program would still assist farmers, but probably not the same farmers. It is not clear what programs of support for agriculture will emerge from the current process of political and legislative wrangling. Still, at the end of the seven-year transition period, the underlying problem of American agriculture—its ability to overproduce and swamp the market—will not have changed. It makes sense to think now about possible interventions to assist agriculture without returning to the commodity price-support programs of the past.

We are at a unique point in the development of both farm- and food-assistance policy. In this time of change, we can build a coalition of farm and food sector interests, antipoverty advocates, and nutrition advocates, and even expand the coalition to include advocates for local and sustainable agriculture, to promote expansion in food-assistance programs that would capture some benefits now provided to the farm sector and use them to benefit the economically disadvantaged also.

My vision is of a nation that recognizes both the wisdom and the moral imperative of assuring adequate food for the needy and vulnerable by means of targeted programs to put food-purchasing power in the hands of the poor. Only a few years ago, nutrition advocates expected the policy debates at the federal level to focus on questions such as whether to make WIC an entitlement program. In the current climate, allocating new funds seems far-fetched even for a program such as WIC, which is widely recognized as effective and is linked to mothers and babies, a merit group of consumers. However, it is still plausible to argue that funds currently allocated to benefit the farm sector could continue to benefit that sector through demand expansion while also providing nutritional benefits. The achievement of these combined goals will require the efforts of a coalition that extends beyond the traditional supporters of either farm- or food-assistance programs.

Reference List

Alderman, Harold. 1986. *The Effect of Food Price and Income Changes on the Acquisition of Food by Low Income Households.* Washington, D.C.: International Food Policy Research Institute.

Brooks, N. L., and Donn Reimund. 1989. *Where Do Farm Households Earn their Incomes?* Agriculture Information Bulletin No. 560. Washington, D.C.: Economic Research Service, U.S. Dept. of Agriculture.

Center on Hunger, Poverty, and Nutrition Policy (CHPNP). 1992. Unpublished data. Medford, Mass: CHPNP, Tufts University.

Clancy, Kate. 1993. "Sustainable Agriculture and Domestic Hunger: Rethinking a Link between Production and Consumption." In *Food for the Future: Conditions and Contradictions of Sustainability,* edited by Patricia Allen, 251–93. New York: John Wiley & Sons.

Community Nutrition Institute (CNI). 1995a. *Nutrition Week* 25 (February 24): 3.

———. 1995b. *Nutrition Week* 25 (April 28): 1.

Devaney, Barbara, and Thomas Fraker. 1989. The dietary impacts of the school breakfast program. *American Journal of Agricultural Economics* 71:932–48.

Egbert, Alvin C., and Stephen J. Hiemstra. 1969. Shifting direct government payments from agriculture to poor people: Impacts on food consumption and farm income. *Agricultural Economics Research* 21 (3): 61–69.

Elitzak, Howard. 1993. Food marketing costs rose little in 1992. *Food Review* 16 (3): 28–29. Economic Research Service, U.S. Department of Agriculture.

Federation of American Societies for Experimental Biology (FASEB). 1995. *Third Report on Nutrition Monitoring in the United States: Executive Summary.* Washington, D.C.: Interagency Board for Nutrition Monitoring and Related Research.

Finke, Michael, and Luther Tweeten. 1995. Using markets to induce better diets. Columbus: Department of Agricultural Economics, The Ohio State University. Mimeo.

Finke, Michael, Luther Tweeten, and Wen Chern. 1995. Economic impact of proper diets on farm and marketing resources. Columbus: Department of Agricultural Economics, The Ohio State University. Mimeo.

Food Research and Action Center (FRAC). 1995. *Community Childhood Hunger Identification Project: A Survey of Childhood Hunger in the United States.* Washington, D.C.: FRAC.

Fraker, Thomas M. 1990. *The Effects of Food Stamps on Food Consumption: A Review of the Literature.* Washington, D.C.: Food and Nutrition Service, U.S. Department of Agriculture.

Gordon, A. R., and P. McKinney. 1995. Sources of nutrients in students' diets. School Nutrition Dietary Assessment Study. *American Journal of Clinical Nutrition* 61 (1S): 232s–40s.

Hanes, S., J. Vermeersch, and S. Gale. 1984. The National Evaluation of School

Nutrition Programs: Program impact on dietary intake. *American Journal of Clinical Nutrition* 40:390–413.

Manchester, Alden. 1991. Food spending. *Food Review* 14 (3): 27. Economics Research Service, U.S. Department of Agriculture.

May, Richard, and Kathryn Porter. 1996. *Poverty and Income Trends, 1994.* Washington, D.C.: Center on Budget and Policy Priorities.

Paarlberg, Don. 1980. *Farm and Food Policy: Issues of the 1980s.* Lincoln: University of Nebraska Press.

Poppendeick, Janet. 1995. "Hunger in America: Typification and Response." In *Eating Agendas: Food and Nutrition and Social Problems,* edited by Donna Maurer and Jeffery Sobal, 11–34. New York: Aldine de Gruyter.

Senauer, Ben, and Nathan Young. 1986. The impact of food stamps on food expenditures: Rejection of the traditional model. *American Journal of Agricultural Economics* 68:37–43.

Senauer, Ben, Elaine Asp, and Jean Kinsey. 1991. *Food Trends and the Changing Consumer.* St. Paul: Eagan Press.

U.S. Bureau of the Census. 1995. *Statistical Abstract of the United States, 1994.* Washington, D.C.: U.S. Department of Commerce.

U.S. Department of Agriculture. 1981. *A Time to Choose: Summary Report on the Structure of Agriculture.* Washington, D.C.: U.S. Department of Agriculture.

11

Government Pathways to True Food Security

Kathleen A. Merrigan

The future of U.S. agriculture depends on reinventing government according to three principles: regulation, diversity, and democratic decision making. These principles will help farmers by ensuring market access and environmental stewardship. Most important, government based on these principles will bring about real food security for all Americans. To attain my vision of U.S. agriculture, we must undergo a disruptive period of heavy-handed government reforms, followed by a true partnership between the public and private sectors.

Updating the Concept of Food Security

Congressman Kika de la Garza (D.-Tex., 1965–96), longtime former chair of the House Agriculture Committee, often closed his speeches with a tale designed to emphasize the importance of agriculture: A multibillion dollar submarine operated by the best-trained officers and complete with the latest weapons and technology is sent on a mission. The president calls the submarine commander to find out how long the vessel can remain submerged to protect the nation—a week? a month? a year? The commander replies, "Mr. President, it all depends on how long the food holds out."

The story is a useful reminder that agricultural policy is fundamental to any national security plan. Over a scale of a few decades, what will "food security" mean? Until recently, many people argued that food security was ensured through the control and enhancement of commodity prices. During the Depression era, the government provided price supports to farmers to bolster farm income, alleviate unemployment in rural communities, and guarantee an adequate food supply. This made a great deal of sense, because at the time one in four Americans lived on a farm; therefore, paying farmers was an efficient

155

way of undergirding a flailing economy. Yet such price supports persist today, more than sixty years later, despite the very different economic climate and the fact that less than 2 percent of Americans now live on farms. While price support advocates employ the rhetorical justification of "food security" as they fend off reforms, a spectrum of organizations from conservative to liberal has called for the elimination of commodity programs. Responding in 1996, Congress took what may be the first step in dismantling commodity programs by decoupling subsidy payments and production.

No one argues that it is time to discard the notion of food security along with outdated commodity programs. With the world population expected to grow by one hundred million people a year for the next thirty years, we must never ignore the importance of maintaining production capability. Nor should we ignore the major effect this industry has on the U.S. economy: in 1994, agriculture-related jobs accounted for 17 percent (22 million jobs) of total U.S. employment and 14 percent ($939 billion) of the gross domestic product (Edmonson et al. 1995). Rather, it is time to modernize our approach to food security by broadening it to encompass the complexities of a stable and sustainable food system.

Food security will require affirmative answers to the following questions: Have we adequately safeguarded our environment to ensure that we will always have the necessary soil upon which to grow our food and safe groundwater to drink? Do we have a clear plan to maintain America's ability to feed itself despite increasing engagement in international trade? Can all people obtain healthy food at a reasonable price? Do we have a diversity of farms and crops such that no single pest or virus could significantly harm our food supply? Are our government agriculture programs democratically determined, and do they support a broad section of the population?

Unfortunately, the answer to all of these questions is a resounding "no."

Regulation Is Long Overdue

At the foundation of my vision of a secure food system is a radically different notion of the proper role of government, one that in the short run involves greater regulation. Admittedly, this goes against current trends. Deregulation is the 1990s buzzword as both Congress and the executive branch strive to curtail government programs, expecting volunteerism and the private sector to fill the gaps. Clearly, the mood of the country is to shake government off its back and to devolve many remaining government activities from the federal to the state and local levels.

My call for greater government control clashes with the popular but incorrect perception that the federal government is and always has been intimately involved in controlling agriculture. This has never been the case. In 1862, when the U.S. Department of Agriculture was created, its role was to conduct research, collect statistics, and provide information to farmers. Although the USDA has greatly expanded, its central mission has always been to provide programs that farmers voluntarily participate in, such as technical and disaster assistance, insurance, crop subsidies, and farm credit. The USDA also provides two kinds of "regulatory" assistance welcomed by farmers: market regulatory programs that ensure high prices (e.g., tobacco and peanut allotments) and food quality programs that protect farmers from market disruptions (e.g., meat and egg inspections). Stories of overregulated farmers unable to farm freely are far-fetched. The often-bashed and ominous-sounding "swampbuster" and "sodbuster" programs enacted in the 1985 and 1990 Farm Bills are decried by farm lobbies as burdensome, but even these are regulatory only in that farmers who voluntarily accept crop subsidy payments from Uncle Sam must comply with minimal conservation standards.

Since farmers comprise the USDA's main constituency, its reluctance to regulate is not surprising. But several other federal agencies involved with agriculture also give farmers the "kid glove" treatment. One example is the Environmental Protection Agency in its regulation of pesticides. Although pesticide law is supposed to protect health and safety, many environmental advocates and scientific authorities alike charge that the EPA's regulation of pesticides is inadequate. The Department of the Interior protects endangered species, but contrary to news reports of farmers forced from their land, few farmers have ever experienced restrictions because of endangered species; in the twenty-three years since passage of the Endangered Species Act, only one landowner has been successfully prosecuted for habitat alteration under this law (Judy Pharris, U.S. Fish and Wildlife Service, and Jim Jontz, Endangered Species Coalition, personal communications, 1996). Farmers also complain about federal regulations concerning grazing of private herds on public land. The controversy is not about overregulation, as many farmers want the public to believe, but rather about the price that the Department of the Interior charges farmers for this grazing privilege, even though the price is always well below market rates. Agriculture also has been given special treatment under labor law (Vaupel, Chapter 9). In years past, it was exempted from basic protections such as minimum wage, unemployment insurance, and workmen's compensation, and it still enjoys exemptions from the National Labor Relations Act and portions of the Fair Labor Standards Act, and few states recognize the right of agricultural workers to organize.

Thus, agriculture has escaped most regulatory efforts, while other industries, such as transportation, banking, and telecommunications, for decades have been shaped and dictated by federal regulation. The rationale for government intervention has been to ensure the safety, accessibility, and affordability of vital services. During the early years of industrialization, the Progressive Era, the New Deal period, and most recently in the early 1970s, government regulation was viewed as a necessary good and was liberally applied. As a result, rural communities have access to telephones, the country is crisscrossed by public transportation, prescription drugs are tested, and people confidently place their life savings into a regulated banking system.

Since nothing could be more essential than agriculture, why is it controlled so much less? The primary answer lies in the history of agrarian democracy that permeates our notion of agriculture and colors our views of farmers and their activities (Danbom, Chapter 1). Our literature and history books are replete with romantic stories of hardworking, honest, nature-loving farmers. This imagery has insulated agriculture from most regulatory crusades.

Everyone wants to help farmers. Pollsters remind politicians of this, pointing to surveys showing the public's commitment to "family farmers." But it is no longer helpful—or acceptable—to treat farmers and agriculture as sacred and therefore exempt from government control. Society would benefit from application of two time-tested regulatory efforts in agriculture—land-use regulation and antitrust law.

Breaking Destructive Land-Use Patterns

Relocating Crop Production

Two production patterns must be reversed if the United States is to have enough environmentally suitable land and water to meet our future food needs. First, aggressive federal government intervention is needed to relocate production regions for several crops. Second, crop rotations must be mandated whenever and wherever they can reduce soil erosion and agrichemical use.

The worst environmental problems associated with agricultural production, including loss of soil productivity, decline of air quality, exposure to pesticides, runoff and leaching of nutrients, and destruction of habitat, are concentrated in a band of land that runs across the eastern, southern, and western coastal regions of the United States (Lynch and Smith 1994). Scientists are now able to pinpoint these environmentally endangered lands, but the political process has yet to respond. There is a serious mismatch between these problem areas and those that receive federal conservation benefits, such as under the Conservation and Wetlands Reserve programs. This is not surprising, because conservation programs have never been purely environmental efforts. They were supported by agricultural lobbies seeking effective supply controls and by conservative lawmakers intent on budget cutting. Better targeting of federal conservation programs would require shifting the allocation of conservation benefits from the center of the country to along the coasts.

The concentration of environmental degradation from agriculture suggests a much larger problem than can be cured by better targeting. Policymakers must confront the discomforting reality that certain crops should not be grown in current production regions. For example, while the humid climate in Florida is suitable for fruit and vegetable production, it is also a perfect climate for many pests that are controlled only by heavy applications of fungicides and other toxic chemicals. Although the Florida Everglades has been home to sugarcane production for decades, it is also home to many endangered and sensitive species. There may be no environmentally sensible way to produce many agricultural commodities in Florida. The same goes for the Central Valley of California, where production depends on irrigation involving high environmental costs.

Suggesting radical production shifts may seem ludicrous in the face of a growing private property rights movement that challenges any government restrictions on land use. If we are to safeguard our long-term ability to grow food, however, we must take radical measures. Before dismissing the idea of heavy-handed government intervention, critics should remember that current production patterns are largely the result of government programs that have influenced where, how, and what crops are grown. Without the sugar tariff, for example, Florida sugarcane production would have been eclipsed by Midwestern and Great Plains sugar beet production years ago. Without laws favoring migrant labor, some intensive vegetable production in California would be unprofitable (Vaupel, Chapter 9). Most strikingly, without subsidies, there would be far less irrigation that now consumes precious water to produce such low-value crops as hay, corn, and wheat that could otherwise be grown in regions that do not depend on irrigation.

The concept of relocating production enjoyed some popularity during the New Deal, when it was argued that people owned their land in partnership with society (Lehmann 1995; Danbom, Chapter 1). Calls were prevalent for national land-use planning, including classifying land according to its best use. In 1934, the National Resources Board released a comprehensive national survey of natural resources and advocated the use of government programs such as farm credit and relief payments to secure cooperation on needed land-use reform. Ultimately, these efforts failed for several reasons, including opposition from newspaper and business leaders who feared that such reforms would require population shifts detrimental to their industries. Nevertheless, the vision of the New Dealers is the kind of vision we need today.

It is time to begin the process anew. We must undertake an extensive study of agricultural land like the 1934 National Resources Board assessment, including an analysis of what it would take to shift certain production among regions. We need to ask how the production map should be reformed. What is our degree of flexibility? Will global temperature and precipitation changes require different cropping patterns? What must be done to offset the economic consequences for regions that will suffer from production shifts? We need to develop a series of "what-if" scenarios as we plan the altered land-use patterns that will sustain us.

Once we have a reasoned plan, we will need to engage the political process and determine what regulations the public will support to achieve more sustainable land use. The options range from outright prohibitions on certain production, to tax incentives or disincentives, to subsidies for sustainable choices, to ever-stricter environmental standards that will slowly squeeze out unsustainable production. All these "fixes" already are occurring, often at the state level. But minor fixes are no substitute for a national plan. Consider the consequences if, for example, Florida reduced its production of fruits and vegetables while the Connecticut River Valley gave up farmland to development and thereby eliminated a major production region that could substitute for much of the supply lost from Florida. Clearly, state action cannot substitute for a national plan and a mandate to get the job done.

Monocultures and Rotations

The second land-use pattern that must be broken by federal intervention is the destructive practice of monocropping. An abundant literature documents the severe environmental consequences of monocropping, including the need for heavy applications of pesticides and fertilizers. As a result, crop rotations have become a premise of sustainable agriculture because they increase soil organic matter, replace nitrogen, and break pest cycles.

How do we regulate against monocropping and encourage crop rotations? Several actions could be taken. First, farmers are reluctant to forgo planting profitable crops in favor of sustainable rotations that return less income. To develop stronger markets for rotational crops, we need to encourage different eating patterns by having the USDA promote a varied diet complete with many legumes and other crops that fit into sustainable rotations. Second, we can provide market incentives or subsidies for crops that are beneficial in a good rotational system but for which market prices are unprofitable. Canola, amaranth, vetch, kenaf, and rye are just a few examples of the many crops that would improve rotational practices but need USDA assistance to become profitable on a large scale. Third, we can require farmers to rotate crops as a condition for receiving farm credit services from the government. Fourth, we can severely restrict manure and fertilizer applications to the point at which legume planting is the only reasonable way of adding nitrogen to the soil. Fifth, we can reform commodity and other programs that work against rotations, such as our research system based on a three-year grant cycle, which makes it impossible to plan rotational investigations that last much longer. Finally, we could require rotations outright for all production areas where it makes environmental sense. All these actions would get people, as the popular bumper sticker states, to "stop treating our soil like dirt!"

Eliminating Destructive Market Practices

Agriculture operates as a free market in that government does not control it. But the ever-increasing concentration of power in the food industry makes a mockery of so-called free enterprise, as farmers, businesspeople, and consumers find markets inequitable and difficult to enter. Few deny that the structure of agriculture is becoming increasingly industrialized. The commercials that punctuate political commentary programs on Sunday morning television brazenly boast one company as "supermarket to the world." Farms are getting larger, markets are becoming integrated, and the "little guy" increasingly is finding himself out in the cold.

Nowhere is this more apparent than in the livestock industry, where the message to producers is "get big or get out." Confinement facilities housing three thousand hogs or more, cattle herds numbering in the tens of thousands, and broiler operations of five hundred thousand birds are becoming the industry norm. Small operators find it difficult to secure the services of packing houses and distributors, either because they are dismissed as too small to bother with or because such services are owned by their large competitors. The four largest firms in the meat-packing industry control more than 80 percent of the steer, heifer, and boxed beef market, greatly reducing the flexibility farmers have in selling their herds (Welsh 1996). Large manure lagoons that accompany most large livestock operations create severe pollution problems. Rural communities are disrupted as small businesses close their doors when large operators bypass them in search of corporate-sized traders to fill corporate-sized orders. Concerns arise over the exposure of workers to toxic fumes and the exposure of communities to smells that extend miles from confinement facilities.

The consequences of industry consolidation are beginning to reveal themselves in other areas as well. There is growing concern over the fees that grocers charge food suppliers for shelf space in retail markets. Industry giants get free access to retail shelf space for popular manufactured items, but smaller companies must pay fees for similar access. These fees cost food suppliers some $6 to $9 billion annually, provoking the *Wall Street Journal* to refer to supermarket shelves as the world's most expensive real estate (Gibson 1988). Small suppliers unable to afford these extravagant fees have limited options; in the top twenty-five markets in the United States, three chains control more than 40 percent of supermarket sales (Gibson 1988). Food suppliers understandably are concerned that these fees restrict entry, harm competition, raise prices, and hinder innovation. The consumer ends up paying more for food and has a smaller selection.

Antitrust law has been used to ensure competition across many industries, but it has rarely been used in agriculture. The Packers and Stockyards Act, administered by the USDA, is among the few agricultural laws that include antitrust violations, prohibiting meat packers from manipulating and controlling the market. Unfortunately, this law largely has been ignored by administrators and poorly backed by the courts, leading a 1996 USDA advisory committee and several senators to advocate increased enforcement of USDA's antitrust powers (U.S. Department of Agriculture 1996a). The Justice Department, Federal Trade Commission, and USDA should be directed to ensure that agriculture has a multitude of competitive markets. I concur with those who argue that rather than promoting a global supermarket run by multinational corporations, the role of government should be to ensure creation of a globe of agricultural villages.

Diversity Is Essential

Diversity is necessary for sound and sustainable agriculture, but U.S. agricultural policy has been based on monolithic thinking that in turn has produced a monoculture system vulnerable to many disasters.

Government should be promoting diversity in several areas. First, diversity is needed in our crops and livestock to protect them from pest and disease outbreaks and to provide the genetic stock that is so important to the development of new varieties. In 1970 a blight destroyed 15 percent of the U.S. corn crop because of the crop's genetic uniformity. Since then, there have been several close calls; in 1996 scientists were fearfully monitoring a root pest of grape vines making its way through the Napa and Sonoma valleys in California, where 70 percent of wine grapes come from the same rootstock (National Research Council 1993).

Second, we need diversity of land. Farms are no longer found in every county. As farms became larger and urban centers grew during the past century, farming became concentrated in several geographical pockets. This has had several ill effects. Concentrated farming, like the lack of crop diversity, makes plants and animals more vulnerable to an ecological disaster, such as drought, that may destroy a region's production capacity. Concentrated livestock operations hurt regional watersheds. Farming has both contributed to and been harmed by the national trend toward one-industry towns. Such towns are hostage to one employer; moreover, in the case of agriculture, they are unable to buy fresh local food.

Third, we need diversity within the ranks of our farmers. Unless government intervenes, most operators of moderate-sized farms will go out of business before the close of the century, leaving a small cadre of full-time farmers who will oversee massive production units. The public has begun to understand the problem of lost farmland but has no similar comprehension of the loss of farming experts who have maintained the nation's fields over decades. With most of our farmers nearing retirement age and too few young people seeing any hope in a farming career, we need to attract new entrants to the farming profession. Otherwise, fewer farmers will mean fewer on-the-ground researchers exploring new techniques, crops, and inputs, many of which could lead to environmental advancements.

Diversity among farmers also means diversity in the technology of their operations. Many crucial on-the-ground researchers may not conform to our traditional notion of a farmer. Some of our best research pioneers, for example, are organic farmers. Yet the system does not foster such diversity. Many organic farmers find it difficult to obtain farm credit because bankers require chemical production methods to ensure high yields. Organic and small-scale farmers complain that contract growing for large enterprises requires them to use certain inputs on a schedule set by the firms that buy their produce. Even the research system is biased against organic production; less than 1 percent of USDA research is on organic methods (Mark Lipson, Organic Farming Research Foundation, personal communication, 1996).

My call for diversity among farmers also includes racial diversity. African American farmers have disproportionately left farming and now account for only 0.9 percent of the nation's farmers (U.S. Bureau of the Census 1994). Their loss means a loss of their unique experiential knowledge of certain crops, preferences, and ecological histories. Also, important ties to communities are lost that could otherwise be used to build community-supported local food systems.

Last, we need diversity in our international trading partners while maintaining U.S. self-sufficiency in food production. In 1995 the United States exported approximately $43 billion in agricultural goods, leaving the net balance of trade for the sector a positive $17 billion per year (U.S. Department of Agriculture 1996b). This statistic sounds good, but it does not reveal the underlying weaknesses in our system. As advocates of the General Agreement on Tariffs and Trade (GATT) and the North American Free Trade Agreement (NAFTA), U.S. policymakers have endorsed the concept of "free trade," but in reality our country has yet to develop a national strategy concerning food imports and exports. We have given little effort to developing domestic value-added agricultural activities instead of exporting raw commodities to other countries that process them and sell them back to the United States at a significant profit. We also must recognize that the lack of controls over imports and exports may leave our food system vulnerable to market attacks or may constrain our ability to act on international human rights violations. During the oil embargo of the 1970s, policymakers became aware of the need for a national oil reserve and began courting a diversity of trading partners who could provide our energy needs. This sort of thinking must also be applied to our food supply so that the United States does not become too dependent on any one country for critical food needs.

What is the government to do about the lack of diversity in all these areas? Several important steps could be taken immediately. First, to secure diversity in crops and livestock, the federal government needs to play a much greater role in preserving genetic material. Our country has preserved seeds for many decades at several sites, most notably at the National Germplasm Repository in Fort Collins, Colorado; these seeds have been used by U.S. farmers, by foreign countries in their efforts to repatriate crops, and by biotechnologists in search of valuable genes. However, the effort is severely underfunded. Second, to secure greater land diversity, the federal government must invest heavily in farmland preservation. This may include paying farmers to maintain farmland, developing a national farmland reserve, or working with state and local governments on zoning laws to protect prime farmland from development (Freedgood, Chapter 6). Third, to maintain a diverse farmer reserve, the government should target benefits such as subsidy programs and credit for small- and moderate-sized farmers to bolster their ability to compete and stay in farming. The government should do more for the historically black land-grant colleges of 1890, which receive far less government support than the 1862 institutions and therefore produce students with far weaker credentials. Finally, to encourage diversity in international trade, we need to assess our trading capacity and develop new markets to prevent the United States from becoming too dependent on any one trading partner.

Democratic Policy Development

All of these measures are feasible only if the relationship between agriculture and government changes radically. The current power structure—the "iron triangle" of agricultural interest groups, agricultural legislative committees, and the administrative agencies of the USDA—excludes broad-based public debate of food security issues. We must overcome problems in the political process if we are to have any hope of achieving a sound agricultural system to pass on to future generations.

Problem 1: Money Politics

In principle, both environmental and commodity group lobbyists have equal access to Congress. But the environmental lobbyist's day begins in the office, whereas the commodity group lobbyist's day is likely to begin at an intimate one thousand dollar fundraising breakfast function with a member of Congress to "informally" discuss pending legislation. Frustrated, many in the environmental lobby wonder whether they would not have much greater effect if they quit their jobs and worked full time on campaign finance reform. Campaign contributions are strongly correlated with congressional decision making, especially in agriculture. Agricultural political action committees contribute millions of dollars to the reelection efforts of House and Senate agriculture committee members (Wirth and Schima 1994; Public Voice 1996). Such findings are discouraging for environmental and sustainable agriculture groups, most of whom are nonprofit organizations unable to compete with industry in the buying of Congress.

The agribusiness lock on Congress is nearly complete. The smoke-filled room is the dominant decision forum, and in those rare instances when issues become public, well-financed interest groups top the list of those invited to testify before Congress to present the "evidence" upon which decisions are made. Without serious campaign finance reform, there is little hope that representatives of public-interest groups will be represented fairly in the corridors of power.

Problem 2: Congressional Committees

Most congressional decisions in agriculture are made in committee, with few amendments made on the House or Senate floor. This means that the agriculture committees control the agriculture debate. This often is a problem because of the narrow membership of the committees. Those who seek membership on the committees usually hail from the few remaining agricultural districts in commodity-producing states. It is unusual for citizens from Los Angeles or New York City, for example, to have a senator or representative sitting on an agriculture committee, whereas the citizens of Bismarck, North Dakota, or Springfield, Illinois, expect to have such a representative. In the Senate, where representation is generally less parochial, the narrow mind-set of the agriculture committee is perpetuated because it is the smallest of all thirteen standing committees, leaving little room for anyone but the staunchest "aggies" to join.

Not surprisingly, the jurisdiction of the agriculture committees also is narrow, allowing them to focus mainly on commodity programs—the overriding concern of their constituents. In 1973 the committees did expand their jurisdiction just enough to encompass nutrition programs, such as Food Stamps. However, this occurred only because the committees needed to link expensive farm bill legislation with nutrition programs to entice urban legislators to vote for these omnibus bills. No effort has since been made to expand the committees' jurisdiction further because their narrow agendas appeal to members who wish to avoid many environmental and consumer issues that challenge the agricultural status quo.

It is time to shake things up in Congress. The agriculture committees should be abolished, with agricultural decision making given to newly constructed committees with broad jurisdiction and diverse membership, or to existing committees, such as those covering natural resources. Already, the labor and human resources committees have jurisdiction over legislation concerning Hatch formula funding for land-grant colleges because it is viewed as an "education" rather than an "agriculture" issue. In the same way, wetlands protection could be transferred to the environment committees, commodity programs to small business committees, "food for peace" aid to foreign affairs committees, and so forth. In this way, per-bushel prices no longer will dominate committee deliberations. Rather, the valid concern over commodity prices will be balanced with related agricultural issues such as inner-city hunger, environmental protection, and rural economic affairs.

Problem 3: The USDA Labyrinth

In 1995 the USDA had 110,000 employees, making it the second largest department in the U.S. government. Budget-conscious policymakers are beginning to question why the USDA should be so large, given the diminishing farm population and advances in communications technology. Budget cutting is a valid objective, but it is not the only reason to examine the USDA's large bureaucracy.

The sheer size of the USDA makes it almost impervious to the influence of sustainable agriculture and environmental groups. The commodity and agribusiness groups have staffs and budgets to canvass USDA employees and follow the details of administrative actions, something the nonprofit sector cannot begin to do. For example, the annual budget of just one large commodity organization exceeds the combined annual budgets of the top ten nonprofit organizations promoting sustainable agriculture. Moreover, years of commodity group and agribusiness lobbying have resulted in a classic case of agency capture. Sustainable agriculture, environmental, and other nonprofit groups simply are unable to meet with enough USDA employees to be competitive in the debate over agricultural policy.

Second, the USDA is so big—with a budget larger than that of several countries—that it has developed its own culture, making change difficult to enact. Safe in the heart of this very large bureaucracy, career personnel are not always quick to respond to mandates from political appointees. It takes a long time for campaign rhetoric to be translated into USDA action, if it is at all.

Finally, the USDA's size has exacerbated the problem of integrating agricultural programs and policies. Collaboration and information sharing among its agencies is poor. Besides being uninformed about the activities of their neighboring agencies, agencies also compete because of overlapping mandates. For example, implementation of the congressionally mandated Water Quality Incentives Program has been crippled by the uncoordinated and sometimes hostile efforts of the

Natural Resources Conservation Service and the Farm Services Agency, two USDA agencies responsible for the program (Higgins 1995).

In the early 1990s much discussion occurred over the need to "downsize" the federal government. At the USDA this translated into a timid phase-out of eleven thousand jobs over seven years and reorganization of several USDA agencies, but no fundamental rethinking has occurred. Veterans of previous administration efforts to "reorganize" view the latest attempt as just another in a series of political exercises to convince voters that Washington is "doing something."

A real solution would require a review of all mission areas and a transfer of many activities to the states and counties as well as the private sector. Furthermore, severe cutbacks in staff are needed to pare the USDA to a reasonable size and allow administrators to reconfigure it to fit current needs. This must not turn into a "last hired/first fired" process but should be an opportunity to infuse the system with new ideas. The USDA is a classic case of an intractable bureaucracy. In large organizations, new staff members arrive with fresh ideas and outlooks, but often, years pass before they are promoted to powerful positions in which they can direct change. The result is that either they have long ago discarded those new ideas to gain acceptance in the organizational culture or the once-new ideas no longer are relevant. Clearly, the USDA's staff must be reduced strategically. Civil Service protections insulate many workers from dismissal but do not prevent agency reorganizations that require staff dismissals, worker buyouts, or shifts of people within the bureaucracy. Concurrently, new procedures must be established that reward innovative thinking with the aim of changing a USDA culture that is open to too narrow a range of interests.

Opportunity 1: Democratize Information

The agricultural iron triangle that determines the rules governing our food supply has too few decision makers and very little diversity. Not only do we need to change political processes as I have just outlined, but we also must develop new procedures and programs to democratize information, expand the scope of interested parties, and share decision making between the "aggies" and "non-aggies."

Too much about agriculture is cloaked in secrecy. Public-interest groups find it difficult to obtain basic information necessary to participate fully in the formulation of agricultural policy. Too much information crucial for public understanding of health and safety issues is classified as confidential business information and therefore is unavailable for public inspection. This includes much of the pesticide information held by the EPA and knowledge about biotechnology innovations under the purview of the EPA, Food and Drug Administration, and USDA. Government officials should define confidential business information more narrowly and seek to amend the law if challenged in court.

Information cloaking also occurs when USDA personnel are reluctant to share information about decision-making processes and the development of ideas, claiming that everything must be "cleared" through political channels before being discussed publicly. The problem is that by the time something clears, it is too late in the policy development process to influence the outcome. Consequently, public-interest groups often are forced to file requests for information under the Freedom of Information Act (FOIA) to obtain basic agency information. As FOIA requests mount, efforts are underway to place exemptions on the FOIA in agricultural law. Rather than trying to weaken the FOIA, as it appears to be doing, the USDA should focus on developing a healthy exchange with the public by reducing the need for use of the FOIA through greater sharing of information and by expediting its response to FOIA requests.

Opportunity 2: Broaden the Scope of USDA Clientele

Several critics have pointed out that although there are federal, state, and sometimes county departments of agriculture, there are no departments of food. This reflects our narrow idea of what government should be doing in agriculture and our failure to consider adequately the broader concept of a food system. The clientele of USDA programs should be broadened. Currently, the USDA's focus is on the production end of agriculture, with too little attention paid to distribution challenges, market development, hunger prevention, waste recycling, and other postfarm food-system concerns (Clancy, Chapter 4).

Special attention should be given to the many rural community needs not encompassed under current agricultural programs. For example, we already have a strong farm credit system that could, if the laws are changed, be used by more than just the farm community. The USDA could become more involved in the delivery of services such as water and electricity to rural areas and in doing so save rural communities money by taking advantage of economies of scale. The USDA could help deliver technologies to rural areas, by bringing fiber optics to rural hospitals to reduce expensive medevacs to city hospitals, for example. Special attention should also be given to urban centers where hunger and malnutrition persist. Ironically, in times of so-called agricultural "surplus," national leaders panic and seek international markets to absorb oversupplies, although every day thousands of our citizens go hungry. (In Chapter 10, Beatrice Rogers suggests ways to serve both goals at once.)

A new political coalition succeeded in getting the Community Food Security Act passed by Congress in 1996. This represents one of the few efforts to think broadly about food systems. The coalition consists of 125 organizations whose activities range from sustainable agriculture to food banking. Their mission is to promote comprehensive community-based solutions to farming and hunger problems in rural and urban areas. The coalition obtained authorization of $16 million over seven years ($1 million in the first year) in federal matching funds for local community food and farming projects in low-income communities. The intent is to build connections among family farmers, environmentalists, antihunger advocates, and community garden and farmers' market activists. This is the kind of effort that could bring new parties to the agricultural debate.

Opportunity 3: Develop Partnerships

The federal government cannot and should not have sole responsibility for carrying out agricultural policy. As already stated, certain USDA functions could be devolved to the state or county level. For example, most decisions in Extension already are made at the state and county levels. Over the years, the federal role in Extension has been cut back, but not its bureaucracy. Furthermore, the Smith-Lever Act of 1914 determines how the federal government distributes most of Extension's annual budget of approximately $400 million, leaving only a handful of national initiatives for the 193 Extension employees based in Washington, D.C., to oversee (U.S. Department of Agriculture 1994). In reality, the USDA no longer administers Extension, and it is time to dismantle the federal bureaucracy that purports to do the job.

Much of what the federal government does in agriculture would be done better if decisions were made in concert with nongovernmental organizations. Laws and regulations need to be rewritten to require innovative processes that bring together a broad array of private and public stakeholders in shared decision making, as exemplified in two significant programs. The USDA Sustainable Agriculture Research and Education (SARE) Program provides grants for research and Extension projects to advance sustainable agriculture. SARE is unique in that grant decisions are made at the regional level by administrative councils consisting of scientists, farmers, Extension leaders, nutritionists, and others. In addition, SARE administrators make special efforts to hold forums to solicit the views of a wide range of interest groups. Similarly, the National Organic Standards Board (NOSB), established to help develop national standards for organic production, brings outsider views into the heart of USDA decision making. The NOSB has fifteen members, including positions for three environmental activists, three consumer representatives, three farmers, one retailer, three food processors, one scientist, and one representative from a certifying organization. The NOSB is unusual in two ways: private citizens formally share power with the USDA, rather than having just an advisory role, and its membership ensures a balance of power and perspective between traditional agricultural interests and other stakeholder groups. Such public-private partnership should become the basis for much of USDA decision making.

Finally, it is important to stress that government is not the answer for all that ails agriculture. Some of the most exciting activities will be community based; if government has a role, it will be to encourage promising community ventures and support them financially. For example, the growing community-supported agriculture movement is based on partnerships between food producers and consumers in which consumers buy a share of the farmer's production. In designing sensible agricultural policy for our future, we need to ask a series of questions: Is the task performed better by the private or public sector? Is the task something that can be done in partnership between the public and private sectors? What level of government, if any, is best for the job?

A Time To Choose, Still

I have laid out three changes needed to improve government's role in agriculture. First, it is appropriate and desirable for government to take a more active regulatory role, although regulation alone is not the solution. Second, government programs should foster diversity in what is grown and how we grow it; unfortunately, some do just the opposite. Finally, I have suggested several reforms that could make our decision-making processes truly democratic.

The highly publicized 1981 USDA report on farm policy, *A Time to Choose*, summed it up best: "Times of studious, deliberate approaches to the design of a forward looking farm policy, rather than adjustment of the previous statute, have been rare" (U.S. Department of Agriculture 1981, 101). It is time to come up with a thoughtful plan of action. A role of government is to ensure food security for all Americans. Currently, food security is not a reality, only a goal. Our agricultural governing system is dominated by the same old thinking, controlled by the same old people, and manipulated through the same old political maneuvers. To bring about change, we must rethink government's role. Our current agricultural governing system is set up for farmers and shaped by agribusiness. But future food security depends on a governing system for the people and by the people.

Reference List

Edmonson, William, Mindy Petrulis, and Agupi Somivaru. 1995. *Measuring the Economy: Wide Effects of the Farm Sector.* Technical Bulletin 1843. Washington, D.C.: Economic Research Service, U.S. Department of Agriculture.

Gibson, Richard. 1988. Supermarkets demand food firms' payments just to get on the shelf. *Wall Street Journal,* November 1, 1.

Higgins, Elizabeth. 1995. *WQIP: The Unfulfilled Promise.* Washington, D.C.: Sustainable Agriculture Coalition.

Lehmann, Tim. 1995. *Public Values, Private Lands: Farmland Preservation Policy, 1933–1985.* Chapel Hill: University of North Carolina Press.

Lynch, Sarah, and Katherine R. Smith. 1994. *Lean, Mean and Green: Designing Farm Support Programs in a New Era.* Policy Studies Report No. 3. Greenbelt, Md.: Henry A. Wallace Institute for Alternative Agriculture.

National Research Council. 1993. *Managing Global Genetic Resources: Agricultural Crop Issues and Policies.* Washington, D.C.: National Academy Press.

Public Voice. 1996. *Ag Committee Watch,* Vol. 1, No. 1. Washington, D.C.: Public Voice for Food and Health Policy.

U.S. Bureau of the Census. 1994. *1992 Census of Agriculture.* Vol. 1, Part 51: United States Summary and State Data. Washington, D.C.: U.S. Department of Commerce.

U.S. Department of Agriculture. 1981. *A Time to Choose: Summary Report on the Structure of Agriculture.* Washington, D.C.: U.S. Department of Agriculture.

———. 1994. *1995 Budget Explanatory Notes for Committee on Appropriations.* Vol. 1. Washington, D.C.: U.S. Department of Agriculture.

———. 1996a. Concentration in agriculture: A report of the USDA Advisory Committee on Concentration. U.S. Dept. of Agriculture Hearing, U.S. Senate Subcommittee on Research, Nutrition, and General Legislation, June 18, 1996.

———. 1996b. *Agricultural Outlook* (June). AO-230. Washington, D.C.: Economic Research Service, U.S. Department of Agriculture.

Welsh, Rick. 1996. *The Industrial Reorganization of U.S. Agriculture: An Overview and Background Report.* Policy Studies Report No. 6. Greenbelt, Md.: Henry A. Wallace Institute for Alternative Agriculture.

Wirth, Kelsey, and Frank Schima. 1994. *The Pesticide PACs, Campaign Contributions and Pesticide Policy.* Washington, D.C.: Environmental Working Group.

Part III

12

Rediscovering the Public Mission of the Land-Grant University through Cooperative Extension

John M. Gerber

The land-grant university was a great public experiment of the late nineteenth and early twentieth centuries. Born during the turbulent days of the American Civil War, the public university system, partially financed through grants of federal land to the states, contributed to the rapid industrialization of American society and particularly to the development of modern agriculture. The emergence of new agricultural technologies and concurrent understanding of the physiology and genetics of crop plants and animals allowed remarkable innovations in the nation's agricultural production system.

The "successes" of the system, such as increases in production and productivity, increased labor efficiency, and inexpensive and plentiful food, are legend. The "failures," such as environmental degradation, resource depletion, the disruption of rural social systems, and the disconnection of humans from the sacredness of the earth, began to be recognized more recently. Regardless of one's personal perspective on the relative value of these developments, agricultural knowledge generated and advocated by the land-grant universities indisputably contributed to many social and environmental changes within and beyond agriculture. Today, this much-heralded public university system has lost its former dominant position as the primary source of new agricultural knowledge. Lately, both the agricultural and nonagricultural elements of the land-grant system have received much criticism from sources both inside and outside the system.

Today, the land-grant agricultural institution is splintered. Two of its traditional land-grant functions, research and extension, are moving apart, fragmenting, and dispersing as they attempt to serve separate masters without the vision, purpose, or common stories needed to maintain the wholeness of the organization. This fragmentation is seen in the system's emphasis on competition rather than collaboration, the failure of its functions to interact effectively, duplication of some activities and weaknesses of others, repeated failures to have both "sides" represented in planning, and the disorientation and disillusionment of the land-grant scientist and educator.

Here I examine the disconnected nature of agricultural research and extension only. Analogies may be drawn to the many other academic disciplines and higher education programs, where fragmentation also is a problem (Scott and Awbrey 1993). Although this problem has been recognized, most of the experiments with institutional change have attempted to treat only the symptoms of fragmentation, with rhetoric about interdisciplinary programs, tactical rather than strategic planning, and mergers of academic units by administrative decree. These superficial changes will not be effective, as the fundamental problem is a failure of purpose. Only a rediscovery of purpose can provide the necessary foundation for a revitalized public university. The agricultural knowledge enterprise that served as the embryo for today's public university system should be a catalyst in this process of rediscovery and rebirth. Since this process must include a renewed commitment to the public, the Cooperative Extension System is uniquely positioned to lead the transformation of the public university because of its historical focus on serving the public good.

The land-grant system will probably never reemerge as the dominant source of technical knowledge in American society. Nevertheless, as a public learning organization, it should claim a new and more important role during the twenty-first century by reaffirming the vision of its founders and then taking the next courageous step in its own evolution. This step will be a transformation from an institutional commitment to the accumulation of information and the creation of knowledge, to include a new focus on the discovery of wisdom, where wisdom is defined as the capacity to realize what is of value. Some researchers will claim that science must be value free. On the other hand, Joan Nassauer makes a case in Chapter 5 for "intelligent tinkering" influenced by shared values as a means toward ecological health.

Intelligent Tinkering

Public universities should provide a forum for discussions of community and personal values as they affect the development of new knowledge. This new focus will result in social and environmental changes as big as those of the twentieth century but without the concomitant drawbacks. The Cooperative Extension System can help build the public learning organization of the twenty-first century by exploring external and internal connectivity.

Improved External Connectivity through Enhanced Public Engagement

A new commitment to the land-grant university's mission of serving the public good is needed to help the institution find its place in twenty-first century American society. Citizens should be actively engaged in the research and extension programs of their land-grant university as both teachers and learners. Creative means of encouraging citizen participation in setting the public research and extension agenda must be tested and implemented. A new ethos of public engagement and new institutional mechanisms must emerge that nurture the spirit of connectivity between the university and the external community.

Improved Internal Connectivity through the Integration of Research and Extension

The commonly accepted separation of research and extension in different administrative units with distinct functions fragments the mission of traditional colleges of agriculture. An institutional culture and administrative structure that support integration of the land-grant functions are needed to create a holistic system that discovers, organizes, communicates, and uses knowledge in service to the public good.

Finally, the acceptance of disciplinary research as the principal means of discovering and validating knowledge interferes with connectivity among university disciplines and between the institution and the public. There is a need for a broader understanding of the many sources of agricultural knowledge, both experiential and experimental. Connectivity across university functions and between the university and the public would be enhanced by the emergence of a new learning organization that sought information, knowledge, and wisdom through the interactions of the hand, mind, and heart.

The New Learning Organization

If the land-grant university is to have a distinctive role in the twenty-first century, it must develop new ways of working to allow better external connectivity with the public and internal connectivity both between research and extension and among the academic disciplines. A new organizational matrix that supports both a disciplinary orientation and a programmatic orientation should replace the traditional disciplinary hierarchical structure. Whereas today, university work is accomplished within discipline-bound departments, tomorrow, work will be carried out by program-focused, interdisciplinary, and cross-functional teams. In the new learning organization, academic departments will continue to develop technical and scientific knowledge; however, unlike today's land-grant organization, public engagement will be the responsibility of program teams that include users of knowledge as full members (Drucker 1993).

The boundary between knowledge generation and knowledge application must be erased. In the new agricultural learning system, farmers and public scientists will learn together in ways that are sensitive to the whole system (social and environmental), not only its component parts. The learning organization will encourage the kind of systems thinking that allows us to see the complex interrelationships that often underlie our greatest problems (Senge 1990). For example, we will understand how our desire for inexpensive food degrades environmental quality; how our pursuit of private funds to maintain public universities reduces public confidence; and how the current university accountability structure, particularly for tenured faculty, has led us to undervalue our mission of serving the public good. Whole-systems learning will include both empirical knowledge and community values and thus lead us toward a discovery of wisdom.

It is unlikely that these changes will happen until both knowledge users, such as farmers, and public educators and research scientists accept a broader understanding of how knowledge is generated and validated. For example, both scientists and farmers generate knowledge by creating and testing hypotheses. Many scientists believe that farmers' knowledge has little value until it is evaluated by currently accepted scientific methods. Academic fundamentalism, or the refusal of the academy to value any truth that does not conform to its own professional standards, results in reduced communication, lack of respect, and limited trust between two communities that should be working together to create new knowledge and put that knowledge to work. Why does this occur? The answer is found in an exploration of values.

Agricultural researchers value truth. They are trained to discover truth through a process that begins with a hypothesis within a community of like-minded disciplinary scientists. They then test this hypothesis by identifying a measurable objective, such as increased yield or improved product quality, and creating artificial environments in which most variables are kept constant, except those under investigation. They then observe how manipulating a few variables affects the chosen objective. Failure to disprove the original hypothesis results in what is believed to be a global (although temporary) truth. The knowledge resulting is valid within the disciplinary environment. This process, which demands falsification of an abstract hypothesis under controlled conditions, represents a social choice by scientists and reflects a value system that prefers universalistic or global "truth."

Practitioners such as farmers also value truth, but they have learned to discover truth through a different process. It, too, begins with the creation of a hypothesis, but within a community of learners that includes their family and neighbors. They test their ideas, not in a uniform, artificial environment, but within complex, ever-changing agroecosystems. Knowledge in this system is desired not so much for advancing universal understanding as for solving local problems. They discover and validate truth through an intuitive understanding of complex relationships among multiple variables, confidence in their own observations, and the utility of the practical solutions they create. The knowledge they develop has validity within the user environment. This process, too, is a social choice and reflects a value system that prefers particularistic or local "truth."

Thus, farmers and researchers are likely to hold different views on what is true or valid knowledge. Researchers' truth is tested within the controlled and homogeneous context of the academic discipline. Farmers' truth is tested within the messy and heterogeneous context of their local communities. While researchers' knowledge is sure to be technically sound, farmers' knowledge is more likely to be socially fitting. To enhance both the technical and social values of knowledge, a hybrid process of inquiry is needed.

The sustainable agriculture movement has challenged agricultural researchers' monopoly on truth. Some researchers who have risen to this challenge are discovering a way of inquiry called "participatory research," in which both technical and social values are explicit goals. Many believe that participatory research offers a suitable method for developing new knowledge, practices, and products while avoiding negative environmental and social effects of new technologies (Gerber 1992).

Participatory research involves both scientists and knowledge users in the inquiry process. The outcome is not only new knowledge, but empowered and perhaps wiser participants who are more likely to take appropriate action on the new knowledge. In participatory research, the knowledge users are expected to help identify real problems, suggest alternative solutions, test those solutions, and interpret the results. A democratic or at least representative system of discovery and validation of new knowledge might ensue if others who are likely to be directly or indirectly affected by the results, such as consumers and environmentalists, are included as active participants. This would be a significant step toward the emergence of a public university system focused on the creation of community wisdom.

The land-grant system is already experimenting with changes that may result in a rebirth of the system as a learning organization focused on the discovery of wisdom. Some of those changes are

- a redefinition of Extension as academic outreach
- new cross-functional, task-focused teams
- a reconnection with the "customer" of the public system
- a rediscovery of the value of indigenous knowledge
- goal-oriented grants
- public review of research and education programs

These experiments offer some hope that the land-grant system is changing, but the evolution of the learning organization will depend on explicit efforts to enhance external and internal connectivity.

External Connectivity: Serving the Public Good

The focus of the "new" land-grant university must be outside the institution—on the public. Extension has a particularly important role to play in making this connection. Our understanding of how Extension can serve American citizens of today and those not yet born is the key to its future as a public institution.

Most public educators agree that they have an obligation to serve the public. But who is the public, and what is the public good? We could argue that anyone who walks into an Extension office or calls the university on the telephone represents the public. After all, they are all citizens, they vote, they pay taxes, and they expect service. This expectation can easily overwhelm a public organization like Extension. If we present Extension as the place to call with any question, no matter how trivial, it will become known as a trivial organization. There is no need for Extension to provide access to information that is available in the public library. We clearly need some guidelines.

One guideline should be that all Extension programs must serve the public good. Many Extension programs are designed not to serve "the public" but to serve particular publics, or special-interest groups such as farmers, homeowners, 4-H volunteers, the rural poor, inner-city youth, or the elderly. Some land-grant leaders claim that this diversity of interests is necessary for the organization to survive. We're told that Extension must serve the particular interests of a diverse array of new groups to expand its political base of constituents. But over the past twenty or thirty years, despite the increase in the number of groups Extension has tried to serve, the system has experienced declining political support. Perhaps adding yet another special-interest group to a stable of constituencies is not the answer. Perhaps by serving many special interests, Extension has failed to serve the public good. I believe that we must learn to serve our traditional constituents such as farmers in new ways, before we expand our target audiences.

We might begin by viewing constituents as humans rather than as farmers, ranchers, environmentalists, or consumers with specialized interests, since all these groups have common interests that can be thought of as basic human needs. Among these needs are affordable, nutritionally adequate food; affordable clothing and shelter; a livable environment; the means to secure a livelihood; and accessible educational opportunities (Burkhardt 1986).

Colleges of agriculture can address these basic human needs by doing what they do best: providing high-quality research and educational programs in agricultural sustainability, food quality and safety, human nutrition, and environmental quality. Of course, they must do so in a way that also serves the particular needs of the special interests, or else the knowledge generated and shared will not be implemented.

The organizational structure and culture of the university must nurture programs that serve the needs of both particular groups and the public as a whole. Furthermore, accountability mechanisms must ensure that the public is involved in guiding the public investment in agricultural research and education. If the goals of a learning organization are determined with the involvement of nonscientists representing the public interest, the accountability needed to maintain a focus on the public good will be ensured. Public input through increased citizen involvement would give direction to the system while building public trust and perhaps enhancing public support.

External connectivity may be enhanced through advisory committees, public forums, and shared responsibility for goal setting. Lawrence Busch and Gerad Middendorf (Chapter 14) describe creative new ways to connect research with the public such as science shops and the consensus conference process. One way that needs more testing is a system of goal-oriented research and education grants. In this scheme, grant proposals are developed by teams that include representatives from the full range of learning organization functions (basic research, developmental research, adaptive research, education, and practice). Broad socioeconomic goals for the program are developed by stakeholders cooperating with key public or private decision makers to ensure public accountability. Specific objectives are identified by working groups of stakeholders, researchers, and educators organized into interdisciplinary, cross-functional teams to enhance both scientific potential and the likelihood of implementation.

Once goals and objectives are established, proposal teams identify constraints to achieving them. Proposals include both strategic (what to do) and tactical (how to do it) plans for addressing the constraints. This might require basic, developmental, or adaptive research or education and practice, depending on the problem. The proposal teams describe the socioeconomic importance of the project, and evaluators review the potential for the project to achieve the goal. Funds made available for supporting projects are managed by the program or project teams and directed to research or educational activities as needed to remove constraints.

Of course, public accountability is not a new idea. Increasingly, funding agencies require a technology transfer plan as part of a research grant proposal to increase the likelihood of implementation of new knowledge. Examples of funding sources that require an implementation or outreach plan are industry grants, commodity group support, nonprofit foundations, and the U.S. Department of Agriculture's Sustainable Agriculture Research and Education (SARE) Program.

Another means of enhancing public accountability is to ask interested citizens to be involved in grant review. This works well within SARE, but SARE represents a small part of the agricultural research and education budget. Discussions have begun at the USDA on how all agricultural research can be more accountable to the public; however, unless a public review is added to the traditional scientific review process, it is unlikely that true public accountability will be achieved. The review process for all public agricultural research and education funds should require evaluation of grants by citizens and technology users (such as farmers) as well as by scientists. The arrogance expressed by scientists and administrators who don't believe that citizens can add anything of value to the grant-review process simply aggravates the current lack of public faith in the process. Besides better reviews, enhanced communication, and improved trust, the interaction of scientists and citizens in this process will itself build community wisdom. These ideas are consistent with Robert Goodman's call (Chapter

13) for agricultural research grounded in a quality public dialogue about society's needs.

These suggestions should not be seen as "anti-science" or a criticism of basic research. In fact, there is danger in not recognizing the value of a fundamental understanding of the physical and social world in which we live. Following criticism received from the national science establishment in the 1970s, the land-grant system made a major investment in its capacity to conduct basic agricultural research. Any future changes must not undermine the opportunity to reap the benefits of this investment. In responding to internal and external pressures to enhance its service to society, the system must build upon its strengths in basic, developmental, and applied research and education. A problem that must be addressed is the fragmentation of those areas of strength.

Internal Connectivity: The Problem of Fragmentation

The public agricultural research and Extension organization that has evolved over the past 135 years is now dysfunctional, and its future is in jeopardy. The land-grant organization should function as a systemically integrated system for the creation of new knowledge, the dissemination of knowledge, and service to society by helping individual learners, families, businesses, and communities put knowledge to work. Today, the system's components are no longer functionally integrated or coupled in a mutually supportive manner.

Specifically, the agricultural research subsystem is serving the masters of the discipline, striving to achieve a self-defined greatness and faltering in its mission of serving the public good. The research subsystem is driven by disciplinary incentives such as peer review of publications, scientific panel approval of grant projects, internally focused awards and recognition, and the self-evaluative process of tenure review, which screens out diversity and results in a classist hierarchy.

On the other hand, the agricultural Extension subsystem is serving both public and private masters, pursuing solutions for many problems for which it may not have greater expertise than other public or private agencies or businesses. Extension not only must focus on problems that serve the public good, it also should give priority to those that require the engagement of both research and extension. This traditional partnership with research provides it with its particular character and unique place among the many public and private educational agencies. When Extension attempts to address problems that do not necessarily require the research subsystem as a full partner, the particular strength of the land-grant system is not employed. This synergistic relationship is demonstrated by current public programs in integrated pest management. Below, I examine the components of the system and discuss how they should be interrelated.

The public learning organization of the twenty-first century should include all the following components:

- *Basic research:* inquiry conducted to understand the physical, chemical, biological, economic, and social processes in nature and society
- *Developmental research:* inquiry focused on developing new products, practices, or technologies based on an understanding of nature and society
- *Adaptive research:* inquiry in which new products, practices, or technologies are refined and fitted into new situations
- *Education:* inquiry in which knowledge users, researchers, and educators learn together and teach each other about new products, practices, or technologies
- *Practice:* inquiry in which products, practices, or technologies are used and evaluated in new situations

In the learning organization of the twenty-first century, inquiry will be the thread that ties these functional components together. Of course, all these components exist with varying strengths within the current system of land-grant universities. The present dysfunctionality of the system is caused by the way the components are perceived to be interrelated.

Most scientists perceive the university research and development system to be linear, with its various functions "wired in series" (Holt 1991). The linear research and development system is sequential, with knowledge coming from new ideas generated through basic research. The new knowledge is peer-reviewed and published and thus made available to applied researchers. Applied researchers may be in contact with Extension educators, who in turn presumably transfer the information to users as a technology or practice through publications, meetings, or electronic mail. The knowledge users are outside the system, recipients of the fruits of the labor of scientists and educators. In this system the new information originates far from the context in which it is intended to be used. The system is designed to optimize the accumulation of information, whether or not the information is used or useful.

In this model, research is discipline-oriented, where the goals of the inquiry process are identified and evaluated within the academic discipline. However, this disciplinary orientation creates a weak link in the chain from research to application, thus jeopardizing the potential utility of the new knowledge and failing to serve society's interests. In this model, it is difficult for knowledge users to provide input on the original ideas until they have been turned into a finished product. At best, implementation of new ideas as a practice or product is left largely to chance, not design. At worst, researchers produce scientifically valid answers to questions that nobody but themselves has asked.

I am not recommending that we abandon disciplinary value for user value, rather that we should create a more effective system that will achieve both. The currently accepted linear model supports the accumulation of information but not necessarily the productivity of knowledge nor the discovery of wisdom. So what is the solution? A new model for organizing the public research and extension system is needed.

A goal-oriented model with the same functions "wired in parallel" would provide a more efficient and effective system for creating new knowledge that is valued within both the discipline and the user environment. A system wired in parallel might have all the important functions occurring simultaneously in a coordinated fashion. Therefore, implementation (user application) is not a passive result of a long chain of weakly coupled events but an active part of the inquiry process that is designed right into the system.

This system can be goal-oriented if each function is focused on clearly identified and agreed-upon goals. High priorities such as economic growth, environmental quality, and human capacity building will likely be achieved through incremental steps in the parallel model. A research and development system built on this model will employ cross-functional teams, using expertise from many disciplines and knowledge users to tackle critical concerns. This system enhances both the production of disciplinary information and the productivity of user-specific knowledge. Finally, once those citizens directly or indirectly affected by new knowledge or new technologies are included in the development process, a public university better adapted to serve the public good through the discovery of wisdom will become a reality.

Conclusion

The land-grant system must nurture a new spirit of connectivity both within the university and between the university and the external community. We must create a broader understanding of learning as an inquiry process taking many forms and in which research includes both controlled experiments by scientists and value-laden experiences among citizens. For knowledge to have value in this new system it must be placed in the context of both the discipline and the user.

Finally, the new land-grant university will not only encourage the accumulation of information and the application of knowledge; it will focus on the creation of wisdom, which is the community's "capacity to realize what is of value—happiness, health, sanity, friendship, love, freedom, justice, prosperity, democracy, creative endeavor, productive work" (Maxwell 1992). The new public learning organization will be in the "wisdom business"; therefore, it must be based on knowledge but driven by social responsibility and tempered by community values. The development and exchange of information, knowledge, and wisdom in service to the public good will be the mission of the land-grant university of the twenty-first century.

Reference List

Burkhardt, Jeffrey. 1986. "The Value Measure in Public Agricultural Research." In *The Agricultural Enterprise: A System in Transition,* edited by Lawrence Busch and William B. Lacy, 28–38. Boulder: Westview Press.

Drucker, Peter F. 1993. *Post-Capitalist Society.* New York: HarperCollins Publishers.

Gerber, John M. 1992. Farmer participation in research: A model for adaptive research and education. *American Journal of Alternative Agriculture* 7:21–24.

Holt, Donald A. 1991. "Organizational Paradigms of Agricultural Research and Development." In *Proceedings of the 1991 Annual Meeting of the Agricultural Research Institute.* Bethesda: ARI.

Maxwell, Nicholas. 1992. What kind of inquiry can best help us create a good world? *Science, Technology & Human Values* 17:205–27.

Scott, David K., and Susan M. Awbrey. 1993. Transforming scholarship. *Change.* July/August.

Senge, Peter M. 1990. *The Fifth Discipline: The Art and Practice of the Learning Organization.* New York: Doubleday Publishing.

13

Ensuring the Scientific Foundations for Agriculture's Future

Robert M. Goodman

Agricultural research must change in fundamental ways if it is to serve the changing needs of agriculture, the demands of national security, and indeed the future of a human population racing toward 10 billion and more. In the United States, economic and demographic trends, changes in the landscape, and our changing roles in the community of nations all have implications for agricultural research, and they call for reordering its priorities. Legitimate public concerns about health, the safety and reliability of the food supply, and the cost and nutritional quality of foods all dictate a rethinking of the agenda and institutional arrangements for agricultural research.

In my vision, agricultural research is fundamentally reformed to serve the needs of modern society and to be responsive to the public. This vision is based on the following principles:

- Agricultural research must serve the broad public, not just a narrow constituency of agricultural producers, agribusinesses, or those exercising partisan political influence.
- It must be scientifically integrative, grounded in ecology, and committed to the highest scientific quality.
- It must be conducted in an institutional context that ensures accountability but must be managed lightly to allow creativity, risk-taking, and invention.
- It must be made attractive to people capable of the highest levels of creativity and scholarship and should be a central part of the intellectual and educational missions of our universities and colleges.
- It should be supported by funding mechanisms and an infrastructure that are open to all.

I thank Jo Handelsman for reading and rereading previous drafts and for countless discussions that improved the result and made the effort enjoyable.

187

Without a renewed commitment to serving society's needs and a drastic overhaul of its place in our nation's intellectual life, agricultural research will continue its decline into pursuit of its narrow self-prescribed interests, unable to attract the outstanding minds or investment needed to make a field exciting, rewarding, and sustainable. As a result, it will continue to lose the political support so vital for assured public funding. Publicly funded agricultural research is justified if it contributes to agriculture's competitive success, sustainability, and environmental quality. Failure to embrace changing social needs and reverse the erosion of public support for agricultural research will contribute to a long-term erosion of our nation's agricultural preeminence and possibly to serious disruptions with implications for internal stability and international security.

Agricultural Research's Harvest of Criticism, Past and Present

Agricultural research and its institutions have been the target of critics from the beginning (Rossiter 1975). Controversy arose from the views of those espousing religious causes, defending farmers' traditional practices, and resisting increases in taxation.

Contemporary criticism of publicly funded agricultural research is motivated by many of the same concerns. However, it has added new ingredients. For more than twenty years, the *results* of agricultural research have been targets of criticism. Instead of celebrating what they consider successes—powerful pesticides, labor-saving equipment, massive export earnings, preservation technologies, and advances in food manufacturing—the champions of agricultural research instead must defend their successes against a wide range of critics who blame the "successes" for increased industrialization, decreased consumer satisfaction, increased food safety concerns, and decreased connectedness to the food system. Also new is a profound and widespread disillusionment and pessimism within the research community itself about the system they work in. Researchers feel isolated from the public they are supposed to serve and from mainstream academic life. For all these reasons, as we rethink the future of agriculture, it also is fitting to rethink seriously the premises for publicly supported agricultural research.

A Fundamental Re-vision of Agricultural Research

The intellectual scope of agricultural research is vast. It embraces many different organisms and different kinds of interactions among them. It encompasses essentially the entire scope of academic disciplines, from agronomy and zoology to anthropology and zoonotic medicine, and its participants range from highly experienced farmers and ranchers to highly trained microeconomists and molecular biologists. Therefore, to prescribe a fully detailed agenda is not feasible and in any case would be inconsistent with my call for flexibility and responsiveness in the research enterprise. However, it is appropriate to sketch some ideas for stimulating the necessary rethinking about the future of agricultural research. These fall into three categories:

- *New themes for agricultural research.* The organizing principles for future agricultural research themes that will serve the emerging needs of agriculture described elsewhere in this volume must be ecological and integrative. Unmistakable signs exist that humans must replace the exploitative model manifest in our pesticide-dependent, industrial system of agricultural production. The unifying principles of genetics provided agriculture with powerful tools, both intellectual and technological. The emerging principles of ecology, integrated with genetics and wisely used, offer society enormous promise to move toward an agriculture consonant rather than in conflict with environmental quality and sustainability.
- *Engaging an informed public in setting the research agenda.* Others in this volume write about engaging and empowering public participation to create new relationships between agriculture and communities, landscapes, labor, and the economy. The agricultural research community should embrace this process. Rather than proving meddlesome, as many contend, new mechanisms for engaged and informed dialogue with the public about the agricultural research agenda may be the key to future political support for agricultural research. Today, in contrast, public commitment is declining and public understanding is weak.
- *Rethinking the roles and funding of agricultural research institutions.* The United States spends a sizable amount of public money on agricultural research, but the institutions that spend most of that money are isolated and insular. I propose instead a funding system open to all prospective investigators, the only qualifications being their interests and ability, not whether they are at a specifically agricultural college or research institution. I also propose a major reallocation of public research funding to universities, where agriculture can and must be integrated with broader educational and intellectual missions, and a corresponding deemphasis of research funding in the U.S. Department of Agriculture.

New Themes for Agricultural Research

Ecological principles should dominate how we think about agriculture and how we choose research questions and approaches. Many of society's concerns about agriculture arise from an emerging appreciation of the limits of agricultural production systems that are based on exploitation rather than sustainable use of resources. Agricultural production must become more responsive to such concerns, as expressed by many authors in this volume. This places a staggering challenge on agricultural research, as discussed by Orr (1992, 50) in his writing about ecological literacy and the planetary consequences of human activity:

> Several conclusions are beyond contention. First, we are crossing critical planetary thresholds or will soon do so. Second, we are woefully ignorant of the critical causal linkages between complex systems and the effects of human actions. Third, we do not have readily available data about the "vital signs" of the planet comparable, say, to the Dow Jones index. Fourth, most research is still directed toward manipulation of the natural world rather than toward understanding of the effects of doing so or the development of low-impact alternatives.

From an advanced, industrial, and highly sophisticated exploitative agriculture, we must redefine agriculture in its relationships to individuals, producers, local communities, nations, and humanity. Agriculture no longer is only local because of both global commerce and agriculture's consequences for the biosphere. Similarly, it no longer is primarily the concern just of those who produce or sell agricultural goods.

This redefinition will not be easy or free of risks. We will know that we are making progress if some ancient truths are reestablished. Such truths include the fundamental connection of individuals and responsibility for their food supply and the centrality of agriculture in human civilization.

However, we largely lack the knowledge required to move wisely and effectively toward more ecologically based agricultural systems while ensuring the production capacity to meet the needs of the population. Thus, a significant part of the task of designing a new agriculture will fall to agricultural research.

The fundamental unanswered questions about agriculture are ecological. My vision of agricultural research therefore calls for a major shift toward research that will contribute to a new level of ecological integrity in agriculture. The questions deserving study will require greater emphasis on biological cycles, community ecological processes, and energy efficiency. Although we must continue to strengthen research with a traditional disciplinary focus, we also must move much more aggressively and with much greater analytical skill and rigor toward the study of complex systems and an understanding of the ecological principles underlying agriculture. A similar recasting of research questions in the social sciences will likely be needed to underpin new models for rural communities and global economic competitiveness.

The major gaps in our knowledge about the biology of agriculture are ecological. Any vision of ecological integrity for agriculture demands better knowledge of the biotic and abiotic interactions of plants and animals with their natural environment. Promising new approaches now allow us to rigorously study such interactions using research tools from molecular biology to address complex ecological questions.

Our objective must be productive agroecosystems that operate in concert with natural systems rather than simplifying or degrading them. The needed understanding will be both mechanistic and holistic. The complexity of biological cycles in a farming system prevents us from controlling every detail. Therefore, the focus in much agricultural practice and research has been on maximizing simplicity. We have concentrated on the development and application of products and the diagnosis and solution of problems. A more ecological approach, embodying both mechanistic and holistic views of agricultural production, will instead focus on managing naturally occurring cycles and anticipating and avoiding problems. The research underpinning of this concept will likely include discovery, descriptive research, and technology development. It will integrate across disciplines and scales and will be rigorous and creative but also practical and outcome-driven. It will be rich with technology but balanced with management know-how. In saying this I am not rejecting monoculture or advocating a return to pre–twentieth-century practices. On the other hand, neither am I rejecting the potential of polyculture or of nutrient management approaches that may look like the methods used on self-sufficient farms of the past. The key to this vision is to know how systems work and to integrate this knowledge in sophisticated ways that will enhance and enrich rather than exploit and degrade the people and natural resources on which agriculture depends.

Agriculture should be in concert rather than in conflict with the rest of the biotic landscape. This will require better ways of monitoring the consequences of agricultural practices and choosing technologies appropriate for the ecosystems in which they are used. For example, we should know much more about the vast unknown world of microorganisms that inhabit soils, plant surfaces, and the digestive tracts of our livestock. Nutrient cycling, biological controls of pests and pathogens, environmental resilience, and increased biodiversity all are reasons for improving our understanding of the microbes with which we and our crops and livestock share the biotic world.

Our thinking about agriculture should not be limited to a field, farm, or particular cropping system. It must extend beyond the farm to the landscape and to the boundaries between cultivation and other forms of land use. Ecology has much to teach us about the wise management of watersheds, uncultivated and fallow lands, and cultivated lands and crops. A long-term view of agriculture should not be limited to the production of food, feed, and fiber. Renewable sources of alternative energy, chemical feedstocks, industrial raw materials, and even pharmaceuticals will likely one day come from "agriculture." We should be prepared for a time when such opportunities will exist for nations with the know-how, technology, and *understanding*. Again, a holistic and ecological outlook allows one to think about meeting agricultural, environmental, energy, social equity, and even aesthetic requirements that will enrich human life. It holds the hope that we will know enough to detect when something is out of whack before a Silent Spring or a Superfund disaster confronts us.

Engaging an Informed Public in Setting the Research Agenda in Agriculture

There is a powerful lesson in the contrast between our political commitment to biomedical research and to agricultural research. Many educated people cannot give a specific reason for investing tax money in agricultural research. Yet individuals and nations are at risk of catastrophic loss of life as much from starvation as from disease. The public and government have a shocking lack of understanding about the role of agricultural research and more broadly about the factors that determine our food system's resilience and vulnerability. Even in developed countries the link between nutrition and health is a major though often unacknowledged public policy concern. This agricultural illiteracy is the direct result of how our educational system, agricultural institutions, and government have marginalized and isolated agricultural research from other areas of public science such as biomedicine.

We must therefore provide the appropriate feedback in our democratic society so that the public feels part of the agricultural constituency and becomes much more engaged in the dialogue about agriculture. This requires that we incorporate education about agriculture into all aspects of life and throughout the educational system. Such engagement should extend to setting the research agenda and will contribute to making agricultural research more central in our national life.

What goals of broad public appeal could return agricultural research to a central position? One such goal that should be driving the agricultural research agenda is to make agriculture and environmental integrity mutually supportive. Agricultural technology is widely viewed as compromising environmental quality, and agricultural research is now seen, with some justification, as contributing to the problem rather than the solution. Agricultural research along the ecological lines that I am advocating can contribute to resolving this problem.

Manipulation of ecosystems is intrinsic to agriculture. The environment that we today accept as "natural" is an artifact of human intervention. The spread of agriculture around the globe replanted much of the earth's surface with nonnative vegetation. However, this does not and should not make the public complacent about the more recent environmental damage from agriculture through pesticide contamination, habitat destruction, loss of genetic diversity, and soil degradation. Because we understand poorly how ecosystems function, we have caused unnecessary and unwitting ecological damage through our agricultural practices. The agricultural research community has been slow to accept the challenge and opportunity intrinsic to these concerns and instead has largely aligned itself until very recently with those forces in agriculture that defended or denied the negative ecological ramifications of many production practices. Despite promising signs of change, we have a long way to go in fully addressing public concerns about environmental quality in agriculture and recasting the agricultural research agenda in a way that will contribute to solutions and provide alternatives rather than merely mitigating damage caused by continuing existing practices. Few thoughtful people in agricultural research today are blatant apologists for environmental damage, but there is much room for more holistic thinking and strategic redirection in the agricultural research community. Greater engagement of the public will help move the research community in the right directions.

Another issue that should drive public engagement in setting the agricultural research agenda is the need for new global mechanisms for reviewing and adjusting the economic system in which agriculture functions. Some progress has been made in lowering subsidies that skew markets and distort international trade, but we still lack integration of these steps with the need to ensure regional and global environmental quality. Concerns include the best use of land resources, the preservation and characterization of global genetic resources, and management of intellectual property. These questions and many others asked elsewhere in this volume raise significant and challenging interdisciplinary research questions in the social sciences.

We need new leadership to shape wisely and comprehensively the funding priorities, funding mechanisms, and mind-sets of participants in the research enterprise. Such leadership might be exercised by a select but nonelitist group of people representing responsible national groups from nongovernmental organizations and foundations as well as progressive thinkers from the public, government agricultural institutions, higher education, and private corporations. Such a group could increase public awareness of the issues, command the attention of political leaders, propose themes and priorities, keep the discussion going during times of controversy, and maintain a balance among the pluralistic, divided, multifaceted interests that collectively make up U.S. agriculture. In the biomedical arena, several foundations and civic leaders have played such a role in contributing to informed dialogue about research needs. Corporate leadership and the professions also have played key roles in supporting the need for investment in both fundamental discovery and more applied research. Nothing like this has existed for agricultural research.

Rethinking the Roles and Funding of Agricultural Research Institutions

The agricultural research agenda as such has only recently been a focus of criticism. Past and current criticism has focused more on research institutions. Institutional arrangements and funding mechanisms create many of the biggest barriers to the progressive changes described above. Such changes will not occur without appropriate incentives; providing these incentives is one function of our public institutions. Appropriate incentives include funding in reasonable amounts, career paths that are respected by others, and an environment conducive to both personal and professional dignity and creativity. We therefore must look seriously at how agricultural research is organized and how and for what purposes funding is allocated. We need new approaches that will tie agricultural research more closely to national needs. This should be reflected not just in the agenda for research and the process used to set it but in the institutional arrangements for research. We also need an approach that will better integrate agricultural research and education in our nation's universities; this includes rethinking the roles of the federal government.

The public (land-grant) universities that house the state agricultural experiment station (SAES) system and the laboratories operated by the USDA'S Agricultural Research Service (ARS) account for most public expenditures for agricultural research. These institutions have a long history of insularity. In a compelling article published more than two decades ago, André and Jean Mayer (1974, 83) wrote about "Agriculture, the Island Empire" as follows:

> Few scientists think of agriculture as the chief, or the model science. Many do not consider it a science at all. Yet it was the first science—the mother of sciences; it remains the science that makes human life possible; and it may be that before the century is over, the success or failure of Science as a whole will be judged by the success or failure of agriculture.
>
> The present isolation of agriculture in American academic life is a tragedy. Not only does it deprive us of the most useful models of the systems approach to human affairs, but it puts us—and the world—in mortal peril.

From a perspective two decades later in the century, we may be tempted to conclude that the Mayers were wrong. Yet, the dominant problems facing global society today are intimately tied to agricultural issues. These include emerging and reemerging infectious diseases (many of which involve the food supply, agricultural production practices, and other features of the agroecosystem), increasing demands on land use and environmental degradation (including the new uncertainties related to global change), and the uncertain prospects for continuing our successes since the 1970s in increasing agricultural yields. Thus, while the time frame for the Mayers' concern now appears to extend beyond the end of the twentieth century, we should not be complacent about the issues they raised in 1974. The fundamental issues they raised then are valid and even more compelling now. What is the cost of maintaining agriculture as an "island empire"? Should we continue to limit participation in research for agriculture to agricultural researchers (defined by their institutional employment)? Should we continue to maintain the public's astounding ignorance about agriculture by insulating agriculture from mainstream educational, political, scientific, and economic discourse?

I believe that the agricultural research enterprise will become better, stronger, and more vigorous if agriculture is integrated into the scientific mainstream. Rather than keeping agricultural research separate from other research, we should create a coherent research capability that integrates it into the agenda of societal issues that drives public investment in the discovery and application of knowledge. Abandoning insularity is central to recruiting the highest caliber of people into agriculture; reconciling and integrating agricultural concerns with health, environmental, and other societal issues; and dealing with the increasingly interdisciplinary nature of knowledge and its applications.

The arguments apply beyond academia: the continued isolation of agriculture from general federal science and technology policy also is alarming. This isolation allows agriculture to be marginalized in political discourse and contributes to its tendency to resist change. For both its own political well-being and the national interest, agricultural research should be integrated into federal science policy as a whole.

How might this happen? There are two basic issues: the future roles of universities and the USDA's laboratories, and the connection between research and practice—in other words, who will do the site-specific, applied research and development that allows the equitable and sustainable development of useful practices?

Reinventing the University's Role

Central in my vision of the future of agricultural research is maintaining and enhancing the U.S. public higher education system. The key elements of this system, which contributes so profoundly to society's well-being, are the pluralism of institutions and the integration of instruction, research, and public service activities. Another strength is that students can shift from a general education into a professional agricultural field at many stages, even after receiving a doctorate in a mainstream discipline such as chemistry, physics, civil engineering, economics, sociology, genetics, or molecular biology.

This vision is not entirely new. Twenty-five years ago, a compelling report from a broadly based national committee assembled by the National Research Council (1972, 50) wrote:

> The scientific stature of personnel engaged in agricultural research is subject to several determinants including the native ability of those attracted into agriculture, the training they receive, and finally the research atmosphere in which they work. The Committee believes that to produce top flight agricultural scientists there should be little distinction between training in agriculture and training in the basic sciences. Agricultural research needs investigators with minds and training equal to those attracted to any other research area. The important problem is to

make the scientific community and particularly the young investigators aware of the problems and opportunities in agricultural research. Interest of the scientific community in agricultural problems and research opportunities needs to be increased. The . . . *increased location of agricultural research in universities, together with greater integration of agricultural scientific education into that of the basic sciences* should contribute to attracting persons of high ability and providing good training for agricultural research. (emphasis added)

This statement reflects the agricultural colleges' long history of insularity, with separate academic programs and an infrastructure that generally is inaccessible to scientists outside the colleges. Moreover, they often duplicate the curricula offered in other units of the same institution, frequently at a lower academic standard. There are some signs of changes like those recommended by the 1972 NRC report. Disciplinary shifts and financial imperatives in some universities are forcing agricultural research faculty into a more central role in the academic life of their institutions, a trend that we should encourage. Moreover, we should open agriculture to branches of the university that traditionally have not been agricultural. Conversely, we should require the traditionally agricultural branches to participate in the general curriculum at institutionally accepted standards.

In advocating this, I am not advocating elimination of agricultural colleges. I see a continuing role for them for at least two compelling reasons. The first is to ensure a place for agriculture in the academic planning process. The integration that I am calling for requires the agricultural colleges to play a leadership role, along with others, in doing interdisciplinary and disciplinary research, offering coordinated curricula, and meeting other academic needs. Today, many agricultural colleges merely colonize the universities they are part of, contributing little to their academic life. Instead, their faculties should contribute to campus-wide undergraduate and graduate curricula, working alongside colleagues from the liberal arts, humanities, and sciences.

The second major reason for maintaining colleges of agriculture is to provide focused and high-quality professional training. But I suggest several cautions. First, specialization should occur much later than is typical in many fields. Second, we must not overemphasize training in skills at the expense of education. A poorly educated person is likely to end up with out-of-date skills, whereas a well-educated one will likely be able to acquire new skills as needed. Often, people who move into fields for which they are *not* trained are the most innovative, creative, and successful. A final caution is that we not interpret "agriculture" too narrowly and instead seek closer integration with training in other professions, such as business, law, medicine, and engineering.

What will guarantee that the "agricultural" in agricultural research will not be lost in this vision? A crucial role will be played by the leadership of universities and colleges of agriculture and by vocal faculty leaders in ensuring that returning agricultural research and education to a central place in our research universities will give it added strength rather than leading to further deterioration and marginalization. Partly, this will result from the strong and collegial participation I am advocating for agriculture in the life of the broader institution. For example, some universities are building new cross-cutting programs as long-range replacements for discipline-based departments. Fields such as rural sociology, applied economics, and reproductive physiology can benefit from integration into such programs while maintaining an appropriate focus on agriculture.

Another way to ensure an important role for agriculture is intelligent planning of funding mechanisms with sufficient levels and duration of funding to be attractive to the broad research community. New funding from federal agencies in several areas related to agriculture illustrates the power of funding programs to shift the focus of the research community. Ecologists and molecular biologists alike (and at times even in collaboration) have been drawn into environmental research by such initiatives. Development of plant and plant-microbe model systems and their appeal for basic researchers have drawn many from outside the agricultural research community to take up such systems for study. The result in both cases has been an increased scope and broadened intellectual appeal of agricultural research. These examples illustrate nicely how funding, along with intellectual factors, powerfully shapes a research agenda and influences participation.

A traditional counterargument has been that agricultural research will disappear unless we maintain specific institutions devoted to it. I do not find this argument compelling. Many important topics in economics research are covered in general economics departments across the country without carving out a separate department for each topic. Why is agricultural economics special? The question is not whether we need a specialized agricultural economics *department*, but how we integrate agricultural economics *topics* into the broader economics research agenda. Agricultural economics would have greater educational value and be more central to the university if it were integrated into the economics discipline instead of remaining separate and isolated, as is typical today. To take another example, research in plant pathology has become so popular in recent years that in a recent survey of the published literature I found that only a minority of plant pathology articles were from departments of plant pathology. These are not arguments either for or against having a department of agricultural economics or of plant pathology. Colleges will organize themselves into departments for many reasons besides intellectual or scholarly ones. Whatever the choice, the goal should be an environment in which the most academically gifted people will find working on agricultural problems highly attractive and in which they have both respect and the necessary funding and infrastructure to do their best work.

Funding is crucial in making any area of research attractive. The discussion often is about funding mechanisms (competitive grants, formula funds, training grants, career enhancement programs, etc.). More important, though, are the amount and duration of funding and the diversity of funding mechanisms. A field as broad as agriculture must have flexible and diverse funding mechanisms. A crucial need is placement of the funds in capable hands, which is usually considered a strength of competitive grant programs. Equally important is for government, public participants in grant making, and the research community to have a culture that embraces an appropriate and constructive level of accountability regarding the use of the funds. Accountability should focus on the creativity and usefulness of the work, not on mechanical formulas. Methods for achieving accountability should be continuing and should entail interaction between those distributing and those using the funds without being onerous or intrusive. When funds are distributed by an agency or donor to an institution—for example, formula funds from the federal government to the SAES—they often are allocated within the institution by competition, with renewal review and accountability measures. On the other hand, institutional funding sometimes is shared by the faculty without competition. Either approach can yield excellent results.

Rethinking the Federal Role

I envision agricultural research that is integrated with other disciplines, drawing students from mainstream educational tracks, contributing to mainstream education, and linking with local communities, businesses, and farmers in ways that are the major purpose and strength of our pluralistic system of higher education. Yet most federal funds for agricultural research are allocated to support intramural research, largely through the USDA's ARS and the Forest Service. Could these funds be invested better? Do we need government laboratories, separate from universities? For two main reasons, I believe that research funds for the ARS and Forest Service should be radically re-structured and reduced, with a corresponding increase in funding used in an intelligent mixture of mechanisms to support university re-search integrated with the education of undergraduate and graduate students. The first reason has to do with focus and the second with sources of funding needed for implementation.

The major priority for agricultural research in my vision is that it better serve public needs for knowledge on which to base new man-agement practices and new ecologically sound technologies by inte-grating many scientific disciplines spanning the social, biological, and physical sciences. These disciplines and the intellectual drive to forge new cross-disciplinary linkages are the intellectual capital invested in our university system of higher education. If our priorities are to strengthen the linkages among research disciplines and integrate them better with education and public service, the USDA's intramural re-search program is largely superfluous.

The second reason is that substantial funds must be available to at-tract high-caliber people into a revitalized university-based agricultural research and education system. Funds are unlikely to be transferred into agriculture from other priority areas. Moreover, the United States already invests heavily in agricultural research but mainly in the USDA's extensive and expensive ARS and Forest Service research pro-grams. Therefore, the issue is not the total level of funding but how we allocate it. Funds presently allocated to federal laboratories would be better spent in the university system.

The USDA's research organizations have been the subject of many decades of review and constant reorganization. It is time to ask, why bother? Why not instead drastically cut the investment in USDA re-search and correspondingly increase funding for university-based agri-cultural research and education? The focus of the ARS and Forest Ser-vice is narrow compared with the multifaceted roles and integrative possibilities offered by the university system. Also, these agencies offer no expertise in agricultural education for the general population or training for the agricultural professions. In contrast, the pluralistic university system can integrate agricultural research with other disci-plines and with education. Shifting most federal funding to university-based research and associated education and training activities would immediately place agriculture at the center rather than the periphery of academic interest and attention. The result would be to attract ex-cellent students and faculty from other fields.

What, then, would be the future of the USDA's intramural research system? One model might be that of the National Institutes of Health. In the NIH, a small intramural program ensures core programming, while a larger extramural program provides flexibility and ensures risk-taking (the grist of innovation) by supporting the creativity of the broad research community and thereby its health and vitality. At the NIH, intramural and extramural (universities and colleges) research receive 20 and 80 percent of total funds, respectively, whereas at the USDA, the split is 80 and 20 percent (General Accounting Office 1996).

Such a scenario invites the question of why there should be any intramural program at the USDA. Borrowing again from the NIH and some corporate-level research models, a compelling answer is that a well-conceived and well-managed central research organization can be pivotal in enabling its associated bureaucracy to better manage an extensive and responsive extramural funding program.

Today's ARS employs a small cadre of distinguished scientists, mostly housed on university campuses across the country. One appealing model would confine a restructured USDA intramural research program to these truly outstanding investigators. As long as the remaining USDA career scientists are allowed the freedom, flexibility, and long-range view that is the hallmark of the best researchers, such a plan would be consistent with my vision. My own view, though, is that the strongest campus-based ARS units are most closely integrated into their campuses. These investigators face a conflict between central government policies and those of the university. So why not take the lesson from this arrangement, where it has been successful, and *fully* integrate these researchers into the university system?

Another radical model for future USDA research would use a mix of extramural funding, based in part on formula funding to the states and in part on grants like those of the NIH and the National Science Foundation, along with a major commitment of federal funds to a career-development program similar to that of the Howard Hughes Medical Institution. Instead of maintaining an expensive and inflexible system of civil servant researchers in national laboratories and even at universities, the USDA would fund the career development of outstanding university faculty members for five to ten years. Support should also be offered to attract promising young faculty into agricultural research careers.

How Would Site-Specific Applied Research and Development Get Done?

Agriculture requires continual innovation in technology and management systems. Often, this requirement is driven by site-specific or local issues and extends well beyond the usual domain of research. The separate facilities and funding mechanisms that give agricultural research its "island empire" tendencies (Mayer and Mayer 1974) are often justified by agriculture's "special" needs for applied research and development. However, the special agricultural colleges and experimental farms were established in the last century to serve a society with very different needs from those of today. U.S. agriculture now is rich with well-educated and skilled entrepreneurs and outstanding managers. Most successful farmers are multitalented, whatever the size of their operations, and often invent and experiment. This community should be drawn into the research process by having them do more applied research themselves on their own farms. Committing more funds and effort to such research and the growing role and increasing sophistication of private advisers will go a long way toward meeting the needs of agricultural producers. As John Gerber discusses in Chapter 12, a critical role for a revised and reinvigorated Extension Service will be to implement such a vision of farm-based participatory applied research and development. This would decrease the support needed for such research in universities and federal agencies, where it has become increasingly out of date and duplicative.

Conclusions

It is time to take seriously and *act* on the advice of André and Jean Mayer (1974, 94) when they wrote "We need a change, both in states of mind and in institutions, if agriculture is to benefit from the intellectual evaluation it deserves and needs." For agricultural research to meet the varied and complex demands of the coming decades, it needs a flexible institutional framework and sufficient and reliable funding that attracts the best of our life scientists, social scientists, and engineers to work on agricultural problems. The research community must be in constant and substantive dialogue with a broad portion of the public, preferably as part of a larger process in which the public participates in setting goals for agriculture and integrating them with goals concerning health, land use, and infrastructure planning. Agriculture and society as a whole are poorly served by agriculture's traditional isolation from such public policy making. Similarly, we should insist on every opportunity to integrate agricultural research into the broader process of educating our citizens and setting research policies for the public good. Research based on ecological principles and grounded in superior scientific knowledge can contribute to the vitality of agriculture. However, it will do so only if we make major changes in how we conduct the business of research, a challenge we must not shun.

Reference List

General Accounting Office. 1996. *Agricultural Research, Information on Research System and USDA's Priority Setting.* GAO/RCED-96-92. Washington, D.C.: United States General Accounting Office.

Mayer, André, and Jean Mayer. 1974. Agriculture, the island empire. *Daedalus* 103 (3): 83–95.

National Research Council. 1972. *Report of the Committee on Research Advisory to the U.S. Department of Agriculture.* Washington, D.C.: National Research Council. Available as PB-213-338 from the National Technical Information Service, U.S. Department of Commerce, Springfield, VA 22151.

Orr, David. 1992. *Ecological Literacy, Education and the Transition to a Postmodern World.* Albany: State University of New York Press.

Rossiter, Margaret W. 1975. *The Emergence of Agricultural Science.* New Haven: Yale University Press.

14

Democratic Technology Policy for a Rapidly Changing World

Lawrence Busch and Gerad Middendorf

Through most of human history people have been wary of technological change. Our ancestors were well aware of its disruptive consequences and attempted in many ways to limit it. The Enlightenment changed all that. Today, technological change is often taken as an undiluted good that, despite temporary "adjustments," will rain down its benefits upon us. Those who oppose new technologies are often branded as latter-day Luddites.

Yet new technologies are usually both constructive and destructive. Most significantly, they redistribute social goods such as wealth, income, power, status, and prestige. In agriculture, labor-saving technologies have lowered the cost of food and drastically reduced the number of people living and working on farms. The mechanical cotton harvester was developed in the 1940s despite an oversupply of labor and was a major factor in the ensuing widespread loss of jobs and migration of agricultural workers out of cotton regions (Dillingham 1966). More recently, the mechanization of fruit and vegetable harvesting forced some producers out of business, transformed the character of the workforce, redistributed wealth and income, and increased the size of farms (Friedland et al. 1981) while lowering consumer prices. Despite the initial success of DDT in cutting insect losses in agriculture, studies at the time of its widespread use showed residues in human fat tissues and its concentration as it went up the life chain (Dunlap 1981).

New technologies also have implications for race and gender rela-

The research reported here is based in part on work supported by the Michigan Agricultural Experiment Station by the National Science Foundation (grant SBE 92 12928). However, any findings, conclusions, or recommendations are those of the authors and do not necessarily reflect the views of these agencies.

tions and for biodiversity. Most of the farm workers displaced by the cotton harvester were African Americans. The vertically integrated poultry industry eliminated the "pin money" that farm wives earned from the sale of eggs and poultry. Improved seed has led us to abandon old traditional varieties (Busch et al. 1995). In each case, two things happened simultaneously: technologies changed and the behavior of people changed. New technologies and new forms of behavior replaced old technologies and old forms of behavior.

Moreover, new technologies both amplify and reduce (Idhe 1979). That which is amplified presents itself to our senses in all its glory, whereas that which is reduced often passes unnoticed. Hence, the tractor's speed and power amplify our ability to plow our fields, whereas they reduce our knowledge of the details of the soil. The new seed impresses us with its higher yield or greater resistance to pests, whereas the old seed somehow seems less adequate, less important. Some may mourn its loss, but it, and the world associated with it, are gone forever.

Although some changes wrought by technology are undoubtedly good, all raise fundamental ethical issues. Unfortunately, these issues have been largely ignored in the past. For many years land-grant universities, the U.S. Department of Agriculture, and agribusiness corporations avoided consideration of the ethics of technology development, believing that science and the technologies to which it gave birth were somehow outside the bounds of ethical debate.

Recently, this situation has begun to change. The orthodoxy of agricultural science has been challenged on several fronts. First, the proponents of sustainable agriculture have suggested that the goals of increased productivity and production are too limited. They specify too narrowly the paths along which technological change should be directed. From Rachel Carson's (1962) blistering critique of the environmental consequences of pesticide use, to Jim Hightower's (1973) critique of the research agenda-setting process, to more recent critiques such as those of Berry (1977), Callaway and Francis (1993), and Allen (1993), the production ethic has come under severe attack.

Second, the new approaches gathered under the banner of biotechnology have brought forth a new era of designer plants and animals. Much research of the past was at the level of the organism or parts of organisms. Therefore, it was and remains relatively crude in how much it could modify the natural world. In contrast, molecular biology has introduced high levels of precision in modifying plants and animals. But by introducing the concept of design, the question of goals again has been made central to the research process.

Third, the critics of molecular biology in agriculture have raised yet another chorus of challenges to conventional thinking about agriculture (Busch et al. 1991). They have articulated the need for ex ante analysis of new technologies that is concerned with more than merely profit.

Finally, the many critiques of the Green Revolution (Pearse 1980; Anderson and Morrison 1982; Shiva 1991) have challenged the idea that international agricultural research has always served the public good. They have suggested that the quest for increased productivity often has masked the desire for greater control and power.

These debates began as narrow critiques of particular technological paths, but they have evolved into wider debates over social goals for agriculture. They have brought socioeconomic effects to the fore—the so-called fourth criterion (after safety, efficacy, and environmental soundness) for assessing the value of agricultural research (Lacy and Busch 1991). Together, these events have shifted the terms of the debate over agricultural technology from the language of "agricultural adjustment" (Rossemiller 1969) to that of technology choice (Busch 1993). Moreover, this is part of a larger debate involving all areas of our society (Sclove 1995).

In this chapter we argue for values that we believe are essential for making wise technology choices. We then ask what an ethically responsible technology policy might look like. Next, we examine various approaches that have been used to improve the process of technology choice. Finally, we propose several interrelated alternatives for producing better technologies in the future.

What Do We Need to Consider in New Technology Development?

Equity. Equity has long been a value of great concern in the development of new technologies. The Extension Service was established in part with equity considerations in the foreground. Its proponents argued that although capital and labor were organized, farmers were not, putting them at a distinct disadvantage. Also, smaller farmers in remote areas were not using new technologies to their fullest advantage. Extension was intended to remedy that situation. Hindsight shows that Extension founder Seaman Knapp and his followers had a naive view of technology, but there is little doubt that equity was what they had in mind. Moreover, rapid adoption of new technology often reduces or eliminates advantages for early adopters. Rossemiller (1969) extended the concern with equity to include compensation for farm workers displaced by the mechanization of fruit and vegetable harvesting. Such workers were never compensated, largely because there were no mechanisms to do so.

Food security. This important value has three interrelated components: adequacy, access, and availability (Busch and Lacy 1984). Availability refers to the amount of food available at a given time and place. Adequacy refers to the nutritional adequacy of the diet. Access refers to people's ability to gain access to food, directly through farming or indirectly by purchase. All three components are necessary to ensure food security, both domestically and internationally. Recently, food security has been linked to the concept of human rights. Oshaug et al. (1994) have argued that minimal notions of social justice require that the right to food be extended to everyone.

Environmental soundness. Presumably, no new technology should contribute significantly to environmental degradation. To put the matter more positively, new technologies should contribute to improving the environment by enhancing sustainability.

Profitability. In our capitalist society, it is essential that new technologies be profitable for those who use them. Presumably, new technologies at least should maintain existing profits, and preferably increase them. However, since some technologies exacerbate the unequal distribution of wealth, the goal of profitability clearly may conflict with the goal of equity.

Safety and risk. Safety is often measured in terms of risk associated with the use of a particular technology. For most agricultural technologies, trade-offs are recognized between benefits and undesired risks. Moreover, risk is not merely a technical issue to be decided by science and government: it is in some sense implicit in all other values (Thompson, Chapter 2). Furthermore, it is inadequate merely to sum up the risks involved in the development and use of a new technology. Equally important is how those risks are distributed among the various actors associated with a technology, from the initial stages of development to the final stages of consumption.

Quality of life and human dignity. Certain technologies, such as those that eliminate heavy toil, do much to improve the quality of life. On the other hand, some technologies may create meaningless, routine drudgery, demeaning those who use them.

Community. Some technologies increase human individuality, whereas others encourage a sense of community. At the extremes, some technologies demand that users ignore the concerns of their neighbors, whereas others may force cooperation and conformity. On another level, some technologies enhance rural communities, whereas others erode their vitality.

Aesthetics. Although often deprecated—especially with respect to production technologies—aesthetic features also are important. In most forms of engineering, design is important; potential users will often reject the product that looks ugly. In agriculture, the appearance of landscapes is very important (Hiss 1989; Nassauer, Chapter 5). Technologies are more acceptable if they support shared notions of attractive, managed landscapes.

Diversity of enterprise. Because of economies of scale, some technologies require the user to specialize in one activity. Others are better suited to farms with a diversity of enterprises.

Human and animal health. Most agricultural technologies have some implications for human and animal health. Presumably, we do not want new technologies that significantly harm the health of those who use them or of farm animals.

Consent. In democratic societies we believe that people should not be subjected to potential harm without their consent. In research, we are careful to inform human subjects about the study. We require that citizens be informed of potentially dangerous substances in food, such as by warnings on food labels about potential ill health effects. Yet, we often say nothing about how new technologies affect people's ability to maintain their livelihood or quality of life. What constitutes "informed consent" regarding new technologies is a matter of considerable debate.

These are just some of the important values in assessing either the processes or the products of technology development. They do not fit into a simple, additive hierarchy of values. In contrast, neoclassical economics assumes that price subsumes all other values, at least in a "truly competitive" market. Thus, monetary value may be used to combine the multiple values held by different people in the economy. Similarly, Maslow's (1970) hierarchy of needs assumes that there is a linear scale on which all values can be placed. Although such approaches are possible, they fail to take into account the incommensurabilities among differing notions of value.

In contrast, Boltanski and Thévenot (1991) argue that there are different worlds of value, each of which has its own hierarchy. Moreover, they note that these worlds are not fully commensurable. Therefore, what may lead to greatness in one world may be cause for concern in another. Ultimately, this means that no simple and unique optimal decision rule can be developed that will ensure that the common good is served. This can only be accomplished through debate and compromise (Benjamin 1990).

An ethically responsible technology policy cannot arise merely from establishing some set of rules to follow or some checklist of values. Neither a single hierarchy of values in which all actions can be ranked nor the reduction of all questions to issues of price, as in cost-benefit analysis, can adequately deal with ethical questions. Such questions require debate over competing and contradictory values to which weights cannot be assigned objectively. Precisely for this reason, compromise must be at the center of technology policy.

Yet compromise can be fair only if all the parties to the compromise are in similar circumstances. If one party can unduly influence the policy process, the process itself must be deemed unfair, regardless of its results.

This suggests, in turn, that technology policy must be made more democratic, for only democratic institutions can make the process fair, producing outcomes that reflect genuine compromise. More democratic technology choice will not ensure that the outcomes are always satisfactory to all, but it will increase the likelihood that the relevant values are considered.

The Current Situation: Some Attempts at Developing Approaches to Technology Choice in Agriculture

What might a more democratic technology policy look like? We are used to considering democracy as limited to the political, but there have been attempts to extend it to the economic and social spheres. In recent years land-grant colleges and other institutions have attempted to increase participation in technology policy. To some extent these approaches have challenged the notion that only experts can make technological choices. Each has both advantages and disadvantages, and some are more innovative than others. The main ones are described briefly below.

Advisory Committees

Advisory committees have long been associated with agricultural technology development. The USDA has more than one hundred such committees (Battan 1995). However, such groups have tended to represent commodity producers rather than a broad range of clientele. Moreover, as advisory groups, they rarely have budgets, missions, or decision-making power. Although they could be a way to increase participation, administrators often use them as rubber stamps for decisions already made.

The Office of Technology Assessment

Recently, the 104th Congress eliminated the budget for the OTA, a federal congressional agency, and the agency shut down. However, as a mechanism to provide support to members of Congress making science and technology policy decisions, the agency's contribution was regarded as high-quality, cost-efficient, and nonpartisan (Carnegie Commission 1991). The OTA was established in 1972 to provide Congress with "objective analysis of the emerging, difficult, and often highly technical issues of the late 20th century" (Office of Technology Assessment 1985). It had the financial and intellectual resources to provide comprehensive, anticipatory, and long-range analyses of the issues involved with technological change. However, it was limited in the extent to which it could meaningfully involve the many stakeholders (including nonexperts) in shaping technological choices, such as through advisory panels, workshops, and public hearings.

Public Hearings

Public hearings are forums in which interested citizens hear presentations regarding the plans of a government agency. They allow citizens to inform themselves on agency policy, voice their opinions, and potentially influence the direction of policy. However, since hearings are often held only late in the process, after decisions to develop particular technologies have been made, in practice they often fulfill only legal or procedural requirements and serve to legitimate decisions already made (Fiorino 1990). Public hearings could, however, be held early in the process of technology development, when little or no money has been committed, when no strong lobbies have been formed, and when the range of choices is great. Such public hearings would be listening sessions for policymakers. As such, they would be a useful first step in a larger program of participation.

Sustainable Agriculture Research and Education (SARE) Program

The USDA attempted to increase research and education on sustainable agriculture through its SARE programs. Beyond research on the biological and natural resource basis of agriculture, SARE's vision of sustainable agriculture includes protecting the health and safety of people involved in the food system, increasing employment opportunities in agriculture, and promoting the well-being of animals (Cooperative State Research Service and Extension Service 1993). Moreover, it is intended that SARE programs will learn from various stakeholders with interest and experience in low-input practices. Thus, SARE maintains both national and regional advisory councils, which are composed of farmers, ranchers, and representatives of nonprofit organizations, agribusinesses, and state and federal institutions. These councils are responsible for establishing goals and criteria for projects and recommending project funding. These two tasks alone suggest that the advisory councils can shape future research on sustainable agriculture. However, this requires them to be more than simply advisory; so that they are not easily disregarded, they must have a larger role in decision making.

Consensus Conferences

The consensus conference was developed and institutionalized in the mid-1980s by the Danish Board of Technology as a "democratic conference" in which ordinary citizens with diverse backgrounds are involved in the assessment of technology and in which the relationship between social priorities and technology choices is made explicit (Danish Board of Technology 1992). The conference consists of a three-day dialogue between a panel of citizens and another of "experts." The citizen panel formulates a series of questions focused on a particular concern, and members of the expert panel respond, articulating their views on the technical aspects, potential benefits, and implications of the technology. Afterwards, citizen panel members draft a consensus report, which is widely disseminated in the media. Usually, the report is then acted upon by the appropriate legislative body. A potential problem with this approach is the assumption that the lay panel represents a cross section of ordinary citizens. Volunteers are self-selected in that they are interested or motivated enough to respond to newspaper advertisements and to volunteer their time. Such people probably hold strong, and perhaps negative, opinions on the technology.

The Keystone Foundation (1991) has run similar conferences, although they have not had the imprimatur of any governmental organization. Yet they have been remarkably effective in producing consensus where none existed before. This approach clearly warrants further exploration.

Science Shops

Many Dutch universities now operate "science shops" as part of their regular activities. Such shops provide technical assistance to any nonprofit group that requests it. They have the great advantage of providing people with direct access to members of the scientific community, including people who otherwise might not have such access.

Impact Statements

Although widely used in the environmental arena, impact statements have not been used in agricultural technology evaluation. Yet especially in the public sector, requiring researchers to consider the impact of their research could have a laudable effect in sensitizing researchers to the scope of the effects of research and in helping to increase the accuracy of predictions in the long run. Friedland and Kappel (1979) have examined this approach in detail and found that it merits serious consideration.

Conclusions

The approaches described above are only the first steps toward wider participation and democratic control over technology development. Further progress will require rethinking and reorganization on several fronts.

Modifying the Boundaries of Science and Technology

The boundaries of science and technology development need to be modified in three complementary ways:

- The literature of science and technology gives the impression that only those with formal training can improve agriculture. Yet there has been a continuous stream of inventions from practitioners. Farmers and others continue to produce new seed varieties, cultivation methods, management approaches, and machinery (Canine 1995).
- The sharp conceptual separation between science and technology needs to be repaired. Today's science is fully technological in both its practices and its results. On one hand, scientific advances depend totally on measuring, displaying, modifying, and transforming the natural world. On the other hand, much science is driven by technological needs. The quest for better means of insect control, more efficient fertilizer use, and more productive animals and plants all pose profound scientific questions. Thousands of scientists are looking for solutions that can be translated into new production technologies.
- The definition of science and technology needs to be broadened to include activities that occur outside the laboratory but that are fundamental to the scientific enterprise. Public and private funding agencies, instrument manufacturers, and general farm and commodity organizations, among others, must be recognized as essential

parts of science and technology development. Science and technology must be seen both as a form of politics and as a human activity influenced by politics.

Extending Democracy to Technology Choice

The days of innocence are gone. No longer can we claim that technology is neutral. It is a force in human affairs that we can and do control, though hardly democratically. There is an astounding contrast between the heated debates over farm and agricultural policy on one hand and the almost complete lack of debate over new farm and agricultural technologies on the other. Here, the land-grant universities have a special role (Gerber, Chapter 12). They can return to and expand the traditions that made them the model of democratic educational institutions and thereby build a future that not only maintains our democratic institutions but expands the scope of democracy itself. They can extend the "fourth criterion"—the socioeconomic effects of technology—to the United States. They can begin to ask how many Americans should be engaged in farming. Is 1.5 percent enough in light of the values claimed for agriculture?

The innovations we have discussed are not inherently outside the mandate of the land-grant universities. They are just the first steps toward democratic control of technology choice. The land-grant universities need to transform how they develop new technologies and provide technical assistance to make these processes more democratic and more inclusive by drawing not only on their faculties but on the public at large. They need to expand into the institutional sphere their ability to produce innovations. They need to reinvent themselves. They have the legislative authority to do so and the ability to evaluate the results of such an endeavor.

Extending Democracy to the Workplace

Reinventing technology policy cannot be limited to the public sector. Today, most new agricultural technologies are financed and marketed by agribusiness corporations, many of which are multinational and even transnational. They, too, must be democratized. They are creatures of the state, publicly chartered as organizations designed to serve the public interest. In principle, they are responsible to all citizens, not only shareholders. Yet they tend to operate as if only one of the values noted above were relevant: profitability. We need to transform these organizations by extending democracy to the workplace. We need to replace the institutionalized conflicts of collective bargaining with institutionalized cooperation in which management, labor, and local communities all have a say in corporate policy. Clearly, this is a task that transcends the democratization of technology policy and the agricultural sector. Yet, it is essential for the democratization of technology development.

Here, too, the starting points have already been established. Several studies have shown the many social benefits of participation in the workplace (e.g., Pateman 1970; Whyte 1991). Moreover, recent management research shows that increased participation not only has social benefits but also increases corporate efficiency (Byrne 1995). *Fortune* magazine has even hailed what it calls "the new post-heroic leadership" (Huey 1994). This new approach abandons the top-down hierarchy of older leadership styles by introducing participation and delegation of responsibility into the workplace. Ironically, the fears of middle management often block such transformations.

As the new biotechnologies and information technologies provide a wider range of choices for corporations, it becomes easier for them to choose technological pathways that both return a profit and (re)produce other important values. But this also increases the complexity of making good decisions. As such, older styles of leadership that put the entire burden on a single individual are likely to become less effective. Leadership styles that involve all workers in decision making are more likely to lead to decisions that are right for both profitability and other values.

Technology choice must be at the center of the debate about the future of agriculture. The divergent goals that are brought to that debate reflect values that often contradict each other. Yet since these values are not part of a simple hierarchy, no single decision rule can decide the direction of agricultural technology. Moreover, in a world undergoing rapid transformation, a long-term vision of technology development is imperative. Therefore, these decisions must be accomplished through a democratic process of debate and compromise in which all stakeholders have an equal voice.

Reference List

Allen, Patricia, ed. 1993. *Food for the Future: Conditions and Contradictions of Sustainability*. New York: John Wiley & Sons.

Anderson, Robert S., and Barrie M. Morrison. 1982. "Introduction." In *Science, Politics, and the Agricultural Revolution in Asia*, edited by Robert S. Anderson, Paul R. Brass, Edwin Levy, and Barrie M. Morrison, 1–12. Boulder: Westview Press.

Battan, Donna, ed. 1995. *Encyclopedia of Governmental Advisory Organizations*. Detroit: Gale Research.

Benjamin, Martin. 1990. *Splitting the Difference: Compromise and Integrity in Ethics and Politics*. Lawrence: University of Kansas Press.

Berry, Wendell. 1977. *The Unsettling of America: Culture and Agriculture*. San Francisco: Sierra Club Books.

Boltanski, Luc, and Laurent Thévenot. 1991. *De la Justification: Les Economies de la Grandeur*. Paris: Gallimard.

Busch, Lawrence. 1993. "Emerging Issues in Technological Change and Technology Assessment in Agriculture." Agricultural Technology and Family Farm Institute Bulletin No. 93003. Madison: University of Wisconsin.

Busch, Lawrence, and William B. Lacy, eds. 1984. *Food Security in the United States*. Boulder: Westview Press.

Busch, Lawrence, William B. Lacy, Jeffrey Burkhardt, and Laura R. Lacy. 1991. *Plants, Power, and Profit: Social, Economic, and Ethical Consequences of the New Biotechnologies*. Cambridge, Mass.: Basil Blackwell.

Busch, Lawrence, William B. Lacy, Jeffrey Burkhardt, Douglas Hemken, Jubel Moraga-Rojel, Timothy Koponen, and José de Souza Silva. 1995. *Making Nature, Shaping Culture: Plant Biodiversity in Global Context*. Lincoln: University of Nebraska Press.

Byrne, John A. 1995. Management meccas. *Business Week* 3442: 122–33.

Callaway, M. Brett, and Charles Francis, eds. 1993. *Crop Improvement for Sustainable Agriculture*. Lincoln: University of Nebraska Press.

Canine, Craig. 1995. *Dream Reaper: The Story of an Old-Fashion Inventor in the High-Tech, High Stakes World of Modern Agriculture*. New York: Alfred A. Knopf.

Carnegie Commission of Science, Technology, and Government. 1991. *Science, Technology, and Congress: Analysis and the Congressional Support Agencies: A Report of the Carnegie Commission of Science, Technology, and Government*. New York: Carnegie Commission.

Carson, Rachel. 1962. *Silent Spring*. Boston: Houghton Mifflin.

Cooperative State Research Service and Extension Service (CSRS). 1993. *General Guidelines for the Sustainable Agriculture Research and Education Programs*. Washington, D.C.: United States Department of Agriculture.

Danish Board of Technology. 1992. *Technology Assessment in Denmark: A Briefing*. Copenhagen: The Danish Board of Technology.

Dillingham, Harry C. 1966. The mechanical cotton picker, Negro migration, and the integration movement. *Human Organization* 25: 344–51.

Dunlap, Thomas R. 1981. *DDT: Scientists, Citizens, and Public Policy*. Princeton: Princeton University Press.

Fiorino, Daniel J. 1990. Citizen participation and environmental risk: A survey of institutional mechanisms. *Science, Technology, and Human Values* 15 (2): 226–43.

Friedland, William H., and Tim Kappel. 1979. *Production or Perish: Changing the Inequalities of Agricultural Research Priorities*. Santa Cruz: Project on Social Impact Assessment and Values, University of California.

Friedland, William H., Amy E. Barton, and Robert J. Thomas. 1981. *Manufac-

turing Green Gold: Capital, Labor, and Technology in the Lettuce Industry. Cambridge, England: Cambridge University Press.

Hightower, Jim. 1973. *Hard Tomatoes, Hard Times.* Cambridge, Mass.: Schenckman.

Hiss, Tony. 1989. Reflections: Encountering the countryside, parts I and II. *The New Yorker* 21:37–63; 28: 40–69.

Huey, John. 1994. The new post-heroic leadership. *Fortune,* 21 February, 42, 44, 48, 50.

Idhe, Don. 1979. *Technics and Praxis.* Dordrecht: D. Reidel.

Keystone Foundation. 1991. *Final Consensus Report: Global Initiative for the Security and Sustainable Use of Plant Genetic Resources.* Keystone, Colo.: Keystone Center.

Lacy, William B., and Lawrence Busch. 1991. "The Fourth Criterion: Social and Economic Impacts of Agricultural Biotechnology." In *Agricultural Biotechnology at the Crossroads,* edited by June Fessenden MacDonald, 153–68. NABC Report 3. Ithaca, N.Y.: National Agricultural Biotechnology Council.

Maslow, Abraham H. 1970. *Motivation and Personality,* 2nd ed. New York: Harper & Row.

Office of Technology Assessment. 1985. *What OTA Is, What OTA Does, How OTA Works.* Washington, D.C.: Office of Technology Assessment.

Oshaug, Arne, Wenche Barth Eide, and Asbjørn Eide. 1994. Human rights: A normative basis for food and nutrition-relevant policies. *Food Policy* 19 (6): 491–516.

Pateman, Carole. 1970. *Participation and Democratic Theory.* Cambridge, England: Cambridge University Press.

Pearse, Andrew. 1980. *Seeds of Plenty, Seeds of Want.* Oxford: Oxford University Press.

Rossemiller, G. E. 1969. Introduction. In *Fruit and Vegetable Harvest Mechanization: Manpower Implications,* edited by B. F. Cargill and G. E. Rossemiller, 3–5. Report No. 17. East Lansing: Michigan Rural Manpower Center.

Sclove, Richard E. 1995. *Democracy and Technology.* New York: The Guilford Press.

Shiva, Vandana. 1991. *The Violence of the Green Revolution.* London: Zed Books.

Whyte, William Foote. 1991. *Making Mondragon: The Growth and Dynamics of the Worker Cooperative Complex,* 2nd ed. Ithaca, N.Y.: ILR Press.

15

Educating Lifelong Learners for the American Food System

William B. Lacy

The education of the next generation of food-system professionals and consumers is central to any vision of the American food and agriculture system and of the broader society. Given the current rapidly changing social, political, and economic context, the next generation will assume a critical leadership role. For this role they will require a wide range of skills, knowledge, and experience and will need to enhance that knowledge through continuing lifelong learning.

This chapter focuses on the formal and informal learning needed for visionary leadership and how public higher educational institutions, particularly the land-grant colleges of agriculture and human ecology, can meet this need. My vision for this lifelong process includes education in four areas: relationships between humans and the natural environment, relationships among humans, diverse ways of knowing, and practical experience. First, however, I discuss the broader social and economic context that requires lifelong learning by both the professionals and clients of the food system.

Social and Economic Context

The enterprises making up the food system function in a broader environment that is changing dramatically. Many political and social factors will demand far-reaching changes in the education of food-system professionals for the twenty-first century.

First, demographic forces will significantly change who is involved in the U.S. food system: the professionals, their partners, their clientele, and the general workforce. The racial and ethnic composition of the population will increasingly diversify: the Hispanic, African American, and Asian populations are expected to increase significantly, and in several states the white population is projected to be a minority within

219

twenty-five years. Roughly one of ten new hires will be Asian; one of six, African American; and almost one of three, Hispanic. Also, two of every three new hires will be female, with women expected to make up half the U.S. workforce by the year 2000 (Snyder and Edwards 1993).

Concomitantly, a major globalization of our economy will occur. Robert Reich, former U.S. Secretary of Labor, has predicted that this will eliminate national economies as we currently understand the concept, so that the only thing remaining rooted within a nation's borders will be its people (Reich 1992). Soon, the primary assets of each nation will be its citizens' skills and insights. The primary political task will be to cope with the strong forces of the global economy, which bestow even greater wealth on the most skilled, while consigning the less skilled to a declining standard of living (Reich 1992). This scenario places even more emphasis on the next generation, their educational development, and their leadership for a more sustainable food system and just society.

These changes in population and the global economy coincide with a major restructuring of the workplace and a second transformation in the American workforce that will strongly influence the structure of the food system and the education of food-system professionals. Peter Drucker (1994) notes that the first workforce transformation occurred at the turn of the century with the decline of agricultural and domestic jobs and rise of blue-collar industrial jobs. These new jobs generally required skills already possessed by the workforce. However, the transformation entailed massive relocation of workers from rural to urban settings and significantly disrupted families and communities. The second transformation, going on now, is the rise of the service economy, reflected in an increase in service jobs and decline in industrial and manufacturing jobs. Today, 71 percent of the gross national product and 75 percent of all jobs are in service industries (Snyder and Edwards 1993). However, these jobs often pay much less than industrial jobs.

These changes have contributed to a loss of real individual and family income for most of the population, including many food-system professionals. Between 1972 and 1992 real weekly earnings in the United States declined by 20 percent (Snyder and Edwards 1993). Moreover, between 1980 and 1990 median after-tax income for U.S. families declined for the lower three-fifths of our population, whereas the highest fifth experienced a real growth of 27 percent. Not only income but also wealth has been redistributed, giving the highest 20 percent of our society an even greater share. By 1992, the lowest two-fifths of families had 15 percent of the income and 0 percent of the wealth, whereas the highest fifth had 45 percent of the income and 80 percent of the wealth

(Macionis 1995). These changes will significantly affect both the professionals and consumers of the food system regarding food distribution and access.

At the same time, the number of "knowledge workers" who require formal education and continuous lifelong learning has increased markedly. During the labor-intensive industrial era, extending through the 1970s, and during a significant portion of the current service economy era, extending through the remainder of the twentieth century, only about 25 percent of all U.S. jobs required mastery of formal reasoning or information skills (Snyder and Edwards 1993). In the next century, in contrast, 75 percent of U.S. jobs may require formal information-handling skills, including graphic and statistical literacy, systematic thinking, and quantitative estimation and allocation, according to some predictions (Snyder and Edwards 1993). Since such a large portion of our population will require these advanced skills, our secondary schools and universities must address ways to teach them to all students, tailoring education to be compatible with their varied learning styles. Moreover, to achieve the full benefits of information technology, our citizenry must be able to use information to make better decisions not only on their jobs but also in the marketplace, in the voting booth, and elsewhere in their daily lives. Social analysts such as Peter Drucker, Lester Thurow, and Alvin Toffler all foresee a critical role for education in preparing workers for the information age or knowledge society. They estimate that by the end of the century these knowledge workers will comprise one-third of the workforce.

During the last twenty years, American business and industry, including the food system, which generates approximately one-sixth of the U.S. gross domestic product, have invested more than $1 trillion in computer and communication technology, application software, databases, networks, and expert systems. Many of these companies also are undergoing many kinds of organizational changes: redesigning their management systems and organizational structures so that their employees can apply new tools and skills to directly improve their work activity; exploring and experimenting with a new team concept for management to replace the multilayered bureaucratic pyramid; and switching from classroom training to on-the-job learning for all types of work-related training, from remedial to high-tech education. The explosive growth of internships as the fifth year of college reflects the blurred line between education and employment. Institutions of higher education are considering better ways of incorporating experiential learning into their instructional repertoire.

New knowledge and information evolve continuously and rapidly, requiring frequent updating by professionals in the food system. More-over, the typical career pattern in the United States involves frequent changes in responsibilities, jobs, and careers. This, too, will make it essential for these professionals to be lifelong learners.

Our food and fiber system and the policy environment in which it functions have undergone major restructuring, become biologically and managerially intensive (e.g., increased reliance on molecular biology, biotechnology, information technology, and financial management), environmentally sensitive, vertically integrated, and highly concentrated. On the other hand, a small but significant portion of production agriculture is part-time, small-scale, and community-based; seeks niche markets; and limits or eliminates the use of chemicals. Also, environmental concerns involving our agricultural and food system have increased, as a result of several forces: the degradation of land, water, and air; the potential for significant global climate change; and the possible harmful effects of food production practices on environmental sustainability and human health. Environmental regulations and policies are leading to greater emphases on sound land-use practices, nutrient and manure management, integrated pest management, integrated farming systems, whole farm planning, and natural resource and ecosystem management. Moreover, many regard this complex system as increasingly fragile and dependent on the cooperation and partnership of many people, organizations, and nations. These changes in structure and policy will require an educational system that better prepares the professional for complexity and change. Each of these changes will affect the education that food-system professionals require. Consequently, they also will affect the educational institutions that provide that education, especially the colleges of agriculture, food, natural resources, and human ecology.

The Content of Education: A Vision for the Future

A society facing accelerating change and complexity particularly needs professionals and clientele who are both well educated and continuously educated, that is, lifelong learners. Effective education for both professionals and consumers will require significant changes in curricula to ensure that students are engaged in crucial areas of learning. As mentioned above, my vision of an effective educational program entails four areas of formal and informal education: relationships between humans and the natural environment, relationships among humans, diverse ways of knowing, and practical experience.

Humans, Nature, and the Food System

First, education must present and explore the complex relationships between humans and the natural environment, including the study of the biological, chemical, and physical world and the relevant social sciences and humanities. As my colleagues and I have argued elsewhere (Busch et al. 1995), a fundamental aspect of this relationship is that nature is socially constituted. Nature is not "natural" but is made and remade by human beings through a variety of processes that can be described collectively as "culture." Thus, plants, animals, and machines are all manufactured—made by human hands—through culture. Even before agriculture, civilization, science, or art, nature was already the sum of the useful and useless, edible and inedible, and loved and hated. Nature was what human beings made it for their ends.

Different societies and communities, however, constitute nature in different ways. In industrialized societies, science and technology are a major part of culture and play a significant role in reshaping both nature and culture. Regardless of efforts to be value-free, science and technology bring values and interests into the process of making nature. In deliberately constituting nature, science withdraws humans from it as it seeks to control and dominate nature. Ignoring the myriad relationships among them, science has artificially separated three key aspects of culture and nature: food, agriculture, and the environment. Science in its transformation of culture and nature remakes nature to suit human purposes and interests, with consequences for both people and their environments. Each transformation of nature also implies a transformation of us as human beings, and of human relations. The way we change nature reveals the kind of people we are and transforms our social relations.

What kind of nature do we want? Because the way we treat nature is an essential part of the way we treat each other, the nature we want must be humane, caring, and befitting of ourselves as human beings.

Another key relationship is between biodiversity and cultural diversity. Our global food and fiber system ultimately depends on both; the erosion of either threatens its long-term sustainability. However, in the last several decades, global biodiversity has substantially eroded. A series of efforts has been developed to slow this erosion, including an elaborate network of germplasm collections and banks, and ecological parks. But without diverse cultures, with their diverse agricultural production systems, diets, and food preferences, we will not have the varied human systems needed to ensure the biodiversity essential to a sustainable global food system (Busch et al. 1995).

These complex relationships between nature and culture; between biodiversity and cultural diversity; and among food, agriculture, and the environment must be core to the education of our food-system professionals. Education about these relationships should place special emphasis on our food and fiber system. Agriculture has been called a conversation between humans and nature (Gardner et al. 1991). Our education of future food-system professionals and consumers must engage them in this conversation while stressing the need to minimize our talking and maximize our listening. This complex but fragile relationship involves population, land use, production practices, natural resources, science and technology, and our global food economy. A broad appreciation of sustainable food and agriculture systems, sustainable development, and a sustainable global ecosystem is central to this understanding.

Human Relationships

Students must also learn about the relationships of humans to each other, including personal, societal, intercultural, and international relationships. The food system and its complex, interdependent human network provide a logical context for examining these relationships.

This study should begin with a development of a sense of self. The social sciences would be central to these studies and should be complemented by the study of language, art, music, and literature. As Wallace Stegner (1986, 10) observed in his essay *The Sense of Place,* the humanities are essential for an understanding of who we are and our sense of place: "No place is a place until it has had a poet." He also observed that developing a learner's sense of place (a place to pioneer, to grow in, to live and work in, to discover, to create, to see the world from, to share, to play in, to return to, and to remember) should be a key element in this education.

Central to a sense of place is one's understanding and appreciation of the food system and its relationship to community. Growing interest in community-supported agriculture, local farm markets, and regional food networks illustrates these important human relationships (Clancy, Chapter 4). Again, analysis of these relationships will require an interdisciplinary approach involving the humanities and social sciences. Our poetry, literature, drama, and film all have been important for our understanding of the food and agricultural system and our sense of place. Poet and writer Wendell Berry (1984, 28) has written eloquently about these relationships, noting that "the good farmer, like an artist, performs within a pattern; he must do one thing while remembering many others. He must be thoughtful of relationships and connections, always aware of the reciprocity of dependence and influence between the part and the whole. His work may be physical, but its integrity is made by thought."

The food system also is a context for examining important cultural values and rituals, economic principles, policies and practices, international trade, and politics. Ethical and value issues related to access to food and to the international trade and political use of food illustrate the complexity of human relationships nationally and internationally. A curriculum that uses our food system to enhance students' understandings of human relationships might include the ethics of feast and famine, images of agriculture in literature and film, food and culture, the sociology and history of agriculture, and international trade and the politics of food.

Ways of Knowing

Students should study several different approaches to knowledge and discovery. Since the Civil War, the scientific and the technological have dominated our postsecondary education on the food system. This approach breaks complex processes into their smallest parts and searches for the universal laws that govern their mechanisms. These ways of knowing will continue to be instrumental in restructuring our food and fiber system, and a clear understanding of science and technology is essential for any future vision of that system. However, complementary ways of knowing—philosophical, theological, experiential, and intuitive—all should be part of the lifelong learner's education. Students need a course that integrates the knowledge derived from scientific disciplines. In addition, however, we all must rediscover the knowledge implicit in stories, in religious belief systems, and in the arts—knowledge that is crucial but is not captured by scientific disciplines. The local, indigenous experience of producers, food processors, wholesale and retail distributors, and food preparers is equally crucial but resists scientific codification. Finally, the study of knowledge systems themselves, through philosophy, psychology, sociology, and the history of ideas, helps establish a perspective on the process of discovery that can be the foundation for lifelong learning.

Practical Experiences

Finally, practical experiences should enrich and complement the other key areas of learning. These experiences should integrate an individual's knowledge of human relationships with nature and each other, and of ways of knowing. They might include course-related projects and case studies, computer and game simulations, semesters abroad, and summer or semester internships. A recent National Research Council (1996, 64) report recommended that "the colleges of agriculture should require students to take at least one internship from a wide range of creative, mentored internship opportunities representing the diverse career settings for which graduates in food and agricultural sciences are prepared."

A course at Cornell University on international agriculture is an excellent example of such practical experience. During the fall semester, students, assisted by a team of knowledgeable faculty members, familiarize themselves with the agriculture and food system of Honduras. For two weeks in January, the students and faculty members travel together in Honduras, observing and analyzing the diversity of agricultural ecosystems, the social organizations and institutions engaged in the food system, and the range of complex social and political issues surrounding this system. Upon their return to campus, the students complete their analysis, share perspectives, and compile their written reports.

Activities such as these should be focused on the food system and engage people in a wide range of experiences, from production to consumption to recycling. While farm production is an essential component of the food and fiber system, other sectors, including agricultural inputs, natural resources, the environment, food processing, distribution, marketing, retailing, preparation, consumption, and waste management, are equally important. Diverse experiential learning options should be available to students. These could occur on both large and small farms, including both specialized and highly diversified operations. They could include both vertically integrated contract animal production systems and small, horizontally integrated crop and livestock operations. Students could also intern with the providers of services and inputs, such as crop and animal advisors, veterinarians, and fertilizer, seed, machinery, and plant and animal protection companies. Other valuable kinds of internships would involve educators (such as state Cooperative Extension educators); local, regional, and national food and agriculture interest groups; and environmental groups focusing on the food system. These experiences also could include various food-processing firms; large and small cooperatives; home-based niche market processors; and different markets, such as farmers' markets, wholesale urban markets, and local, regional, and national retail food companies. Finally, experiences in the state and national political arenas and the food and agricultural media would enhance students' understanding of the system. These experiences should challenge the students to integrate their understandings of the biological, physical, social, cultural, economic, and ethical dimensions of this complex system.

A capstone course would be an effective way to integrate formal classroom learning with the field experience. Students would be expected to develop presentations that capture the important elements of their experience, share their analysis with fellow students, and prepare a final report that incorporates the suggestions from the class. Finally, they would return to the site of the experience and share their report with their mentors and community members.

These four key areas and approaches must be integrated into curricula, programs, and mechanisms to support effective lifelong learning. This will require an educational setting that promotes and develops the student's ability to think critically, integrate theory with practice, successfully use a range of communication options, and solve problems creatively. Now more than ever, students must have these skills to address the rapid changes that are occurring, to make informed and critical decisions, and to provide visionary leadership. This will entail rethinking where and when education will be conducted and who will be the players.

Changes and Challenges for Land-Grant Universities

Colleges of agriculture and human ecology have been a major force in restructuring the American agricultural system and in educating its professionals and consumers. In the middle of the nineteenth century, U.S. universities were significantly transformed, with two new models for higher education emerging. The first, based on the German university, was primarily devoted to the advancement of knowledge through graduate scholarship and research. The second, complementary model, the American populist land-grant system established in 1862, held that higher education should serve qualified young people from all backgrounds. Central to that model was Senator Justin Morrill's proposal to establish public colleges for agriculture and the mechanic arts in every state. In the ensuing century and a third, these land-grant universities and their colleges of agriculture have been highly successful in creating new knowledge and in using it to educate the students, scientists, and practitioners of our agriculture and food system. Today, however, they face the need for fundamental, far-reaching restructuring to meet the educational requirements of the twenty-first century's food system and to educate the future leaders who will shape that system.

Today, public universities are being asked to address social issues that seem intractable for the larger society (Gerber, Chapter 12). For example, many economic development policies, from national down to local, assume that appropriate partnerships with universities will achieve scientific and technological breakthroughs, create jobs, and fuel economic development. Financial and political pressures encourage closer university-industry ties and the commercialization of university research through the new products and processes. This has been a long-standing tradition in colleges of agriculture, but large transnational companies with science-based product lines have significantly deepened the relationship, reigniting the debate about the wisdom of such associations. Will these links entangle faculty members in intellectual and financial conflicts? Will universities be diverted from their primary functions of teaching and basic research?

Also, the portion of the population traditionally regarded as college-age will continue to decline in actual numbers and proportionately as our population ages and low fertility rates are maintained. During the 1980s, this population declined 15 percent. With real incomes dropping for much of the population and poverty growing (a fifth of our children now live in poverty) and with the average college tuition increasing rapidly (doubling during the 1980s), higher education increasingly is inaccessible to many potential students.

Moreover, public land-grant universities, particularly colleges of agriculture and human ecology, are facing challenges to their traditional way of serving the food and agricultural system and rural communities. A diverse array of potential new clients and partners, including organic and sustainable agricultural producers, environmentalists, and consumer advocates, are requesting a broader educational and research agenda to meet their more diverse needs.

A recent study (Dillman et al. 1995) examined what these heterogeneous interest groups really want from their colleges and universities. This survey, based on a representative national sample of adults, provided compelling evidence that lifelong learning has become a reality for most Americans. Acquiring additional education and training throughout one's adult life already is common. Among people who are not retired, Dillman and his colleagues reported that four-fifths not only think that additional training or education is important for them to be successful in their work but also have received job-related training or education in the last three years. Moreover, almost three-fourths indicated an interest in obtaining a college education in the future and in taking a noncredit college course. More than half reported that they would definitely or probably take a college course for credit in the next three years. Even more remarkable is that regardless of age, income, race, or ethnicity, most Americans want to continue their education well past early adulthood. Still, not everyone is achieving their educational goals. Nor are our educational institutions necessarily willing and able to meet these new demands. Our land-grant colleges of agriculture and human ecology, however, are in a unique position to do so (Gerber, Chapter 12). With their Cooperative Extension systems, these colleges have classrooms in almost every county in every state in the nation. Furthermore, they often have sophisticated communication technology to link each county classroom to the faculty and resources of the main campus. Finally, the Cooperative Extension staff has highly educated professionals with advanced degrees and strong practical experience. In cooperation with campus-based educators, they have worked successfully on the regional and national levels for decades and offer a unique but still untapped potential.

Because of the changing social and economic context I discussed above and the emerging needs for continuing and lifelong learning, higher education must address several conflicting challenges: improving and broadening undergraduate and other postsecondary education; achieving a better balance at the university level between teaching and research; developing distance education systems for both formal and informal instruction and learning; planning for lifelong, learner-centered, interdisciplinary learning; globalizing campus programs in the classroom, in the laboratory, and within the broader university community; and restructuring both academic and administrative units to address the tough financial problems facing the institutions and potential students of higher education.

To address these challenges, our institutions of higher education, particularly the land-grant colleges of agriculture and human ecology, must reexamine the central concept of scholarship. Research has been preeminent in our colleges of agriculture and human ecology, but in the recent Carnegie Foundation report *Scholarship Reconsidered,* Ernest Boyer (1990) suggested that we must adopt a new paradigm of scholarship, one that encompasses and promotes not only the discovery of knowledge but also its integration, communication, and application.

Conclusion

Successful implementation of this vision will require aggressive strategic planning, creative and innovative action, and a strong commitment to excellence by colleges of agriculture and human ecology and other institutions of higher education. It will require integrating strong disciplinary scholarship with carefully developed interdisciplinary systems approaches. Higher education will need to explore and use a variety of techniques, technologies, and strategies. These efforts will require an understanding of the backgrounds, experiences, motivations, and values of students as well as insights into their learning styles and preferences. An overriding goal should be to educate students, both potential professionals and clientele of the food and agriculture system, as critical thinkers, problem solvers, and continuous lifelong learners. As Bob Goodman (1992, 49) wrote in *Agriculture and the Undergraduate,* "The students needed most in our changing world are those most likely to think globally, to act creatively, to value diversity, to behave responsibly, to respond flexibly, and to interact cooperatively."

Reference List

Berry, Wendell. 1984. "Whose Head is the Farmer Using? Whose Head is Using the Farmer?" In *Meeting the Expectations of the Land,* edited by W. Jackson, W. Berry, and B. Colman, 19–30. San Francisco: North Point Press.

Boyer, Ernest L. 1990. *Scholarship Reconsidered: Priorities of the Professoriate.* Princeton: The Carnegie Foundation for the Advancement of Teaching.

Busch, L., W. B. Lacy, J. Burkhardt, D. Hemken, J. Moraga-Rojel, T. Koponen, and J. de Souza Silva. 1995. *Making Nature, Shaping Culture: Plant Biodiversity in Global Context.* Lincoln: University of Nebraska Press.

Dillman, D., J. Christenson, P. Salant, and P. Warner. 1995. *What the Public Wants from Higher Education: Workforce Implications from a 1995 National Survey.* SESRC Technical Report #95-52. Pullman: Social and Economic Sciences Research Center, Washington State University.

Drucker, Peter. 1994. "The Age of Social Transformation." *Atlantic Monthly,* November, 53–50.

Gardner, J. C., V. L. Anderson, B. G. Schatz, P. M. Carr, and S. J. Guldan. 1991. "Overview of Current Sustainable Agricultural Research." In *Sustainable Agriculture Research and Education in the Field,* 77–91. Washington, D.C.: National Academy Press.

Goodman, Robert. 1992. "The Challenges for Professional Education in Agriculture: A Corporate Vantage Point." In *Agriculture and the Undergraduate: Proceedings,* 41–51. National Research Council. Washington, D.C.: National Academy Press.

Macionis, J. J. 1995. Society: The Basics. 3rd ed. Englewood Cliffs, N.J.: Prentice Hall.

National Research Council. 1996. *Colleges of Agriculture at the Land Grant Universities: Public Service and Public Policy.* Washington, D.C.: National Academy Press.

Reich, Robert B. 1992. *The Work of Nations.* New York: Vintage Books.

Snyder, D. P., and G. Edwards. 1993. America in the 1990's: An economy in transition, a society under stress. Paper presented at the Experiment Station Committee on Organization and Policy Futuring Conference, June, Washington, D.C.

Stegner, Wallace. 1986. *The Sense of Place.* Madison: Silver Buckle Press.

Contributors

Peggy Barlett is professor of anthropology at Emory University. Previously she was in the sociology and anthropology department of Carleton College. Her most recent book, *American Dreams, Rural Realities: Farm Families in Crisis,* shows the relationship between farm family roles and the farm's financial survival under severe economic conditions. She has also done extensive research on agrarian communities and farmers' decision making in Latin America, relating them to large-scale economic and ecological changes. Barlett is past president of the Society for Economic Anthropology.

Lawrence Busch is professor of sociology at Michigan State University. His research focuses on agricultural research, extension, and higher education policy both in the United States and abroad. He recently coauthored *Making Nature, Shaping Culture: Plant Biodiversity in Global Context* and has begun a volume on the dilemmas of development. Busch is a fellow of the American Association for the Advancement of Science and serves on the scientific advisory board for CIRAD, the French foreign agronomic research agency.

Kate Clancy is director of a national public policy and visioning project at the Henry A. Wallace Institute for Alternative Agriculture. Formerly she was professor of human nutrition at Syracuse University and nutrition policy advisor at the Federal Trade Commission. She has written about and spoken on a variety of food-system issues and policies over the past twenty years. Clancy is past president of the Agriculture, Food, and Human Values Society and the Society for Nutrition Education and currently is a member of the Food Advisory Committee of the Food and Drug Administration.

David B. Danbom is professor of history at North Dakota State University, where he has taught since 1974. He has published extensively

231

on the history of American agriculture, rural life, and agricultural science. His books include *The Resisted Revolution: Urban America and the Industrialization of Agriculture, 1900–1930*; *"The World of Hope": Progressives and the Struggle for an Ethical Public Life*; and *Born in the Country: A History of Rural America*. He also wrote the centennial history of the North Dakota Agricultural Experiment Station, titled *Our Purpose Is To Serve*. Danbom is past president of the Agricultural History Society.

Julia Freedgood is director of Farmland Advisory Services for the American Farmland Trust, where her activities have included heading the nationally recognized study "Does Farmland Protection Pay?", editing a newsletter for farmland conservation professionals, and producing the video documentary "Farmland Forever." Previously she was executive director of the Federation of Massachusetts Farmers' Markets. Freedgood co-organized the conference "Environmental Enhancement through Agriculture" (November 1995) and is a member of the Sustainable Agriculture Research and Education Administrative Council.

John M. Gerber is director of Extension and professor of plant and soil sciences at the University of Massachusetts. Previously he was assistant director of the Illinois Agricultural Experiment Station and program leader of the Illinois Cooperative Extension Service. His research has focused on the physiology and ecology of vegetable production systems. Gerber has written and lectured widely on participatory research and education, the integration of research and Extension education, and the role of citizen input in land-grant universities. He is a former vice president of the American Society for Horticultural Science.

Robert M. Goodman is professor at the University of Wisconsin–Madison, where he is a member of the department of plant pathology, the graduate program in cellular and molecular biology, and the biotechnology training program. Previously he was senior scholar-in-residence at the National Research Council, executive vice president at Calgene, Inc., and a professor in the department of plant pathology and the international soybean program at the University of Illinois at Urbana–Champaign. He works on the molecular regulation of plant defense genes and interactions of plants with noninvasive, beneficial microorganisms.

William B. Lacy is director of Cooperative Extension, associate dean for the College of Agriculture and Life Sciences and the College of Human Ecology, and professor of rural sociology at Cornell University. He was previously assistant dean for research and professor of rural sociology at Pennsylvania State University and is past president of the Agriculture, Food, and Human Values Society. His research focuses on science policy, agricultural research, biotechnology, and sustainable agriculture. Lacy recently coauthored two books: *Plants, Power and Profit: Social, Economic and Ethical Consequences of the New Biotechnologies,* and *Making Nature, Shaping Culture: Plant Biodiversity in Global Context.*

Mark B. Lapping is professor of public policy and provost and vice president for academic affairs at the University of Southern Maine. He was founding dean of the Bloustein School of Planning and Public Policy and associate director of the New Jersey Agricultural Experiment Station at Rutgers University, a dean at Kansas State University, and founding director of the School of Rural Planning and Development at the University of Guelph in Ontario. Lapping is the author or coauthor of several books, including: *Rural Planning and Development in the United States; The Small Town Planning Handbook; A Long, Deep Furrow: Three Centuries of Farming in New England;* and *Rural America: Legacy and Change.*

Kathleen A. Merrigan is senior analyst at the Henry A. Wallace Institute for Alternative Agriculture. Before that she was senior staff member for the U.S. Senate Committee on Agriculture. Merrigan has worked at the Texas Department of Agriculture. She is a member of the National Organic Standards Board and the board of the Organic Farming Research Foundation and is involved with activities of the National Campaign for Sustainable Agriculture, the World Resources Institute, and the Land Institute. Currently she is a Ph.D. candidate in environmental planning and policy at Massachusetts Institute of Technology.

Gerad Middendorf is a Ph.D. candidate in the sociology department at Michigan State University. He is coauthor of several articles on biotechnology and agricultural cooperatives. His interests include the extension of democratic theory to international development policy issues.

Joan Iverson Nassauer is professor of landscape architecture at the University of Minnesota. She is a fellow of the American Society of Landscape Architects and international vice president for policy of the International Association for Landscape Ecology. She formerly was on the faculties of the University of Illinois and Iowa State University. Nassauer's work concerns the perception of nature in agricultural and urban landscapes and how people's values can be used in ecological design and planning. Her research has been used by the Natural

Resource Conservation Service to improve public acceptance of novel conservation techniques.

Max J. Pfeffer is associate professor of rural sociology at Cornell University. Before joining the faculty at Cornell he was on the human ecology faculty at Rutgers University, where he undertook research on labor markets, farm workers, and metropolitan agriculture. Pfeffer was director of the Metropolitan Agriculture in the Northeast project supported by the Northeast Regional Center for Rural Development.

Beatrice L. Rogers is dean for academic affairs and professor at the Tufts University School of Nutrition Science and Policy. An economist specializing in food and nutrition policy in both the United States and developing countries, Rogers's research is concerned with how food price policy and household income affect food consumption and household food security. In the United States she has studied the Food Stamp Program and the measurement of poverty and hunger. In developing countries she has analyzed the effects of agricultural policies, food aid, and social safety net programs. She has been a consultant to the U.S. Department of Agriculture, USAID, the World Bank, and the United Nations.

Karl Stauber is the president of the Northwest Area Foundation in St. Paul, Minnesota, which makes grants focused on economic and sustainable development in the Northern Plains and Pacific Northwest. Stauber was the first undersecretary for research, education, and economics at the U.S. Department of Agriculture, overseeing the integration of the department's physical, biological, social, and statistical science agencies. He also led the Clinton Administration's efforts to reform and update the Research Title of the Farm Bill. Stauber holds a Ph.D. from the Union Institute in Cincinnati, Ohio.

Paul B. Thompson holds the Joyce and Edward E. Brewer Chair in Applied Ethics at Purdue University. He formerly was professor of philosophy and agricultural economics and director of the Center for Science and Technology Policy and Ethics at Texas A&M University. He has written or contributed to five books, including *Sacred Cows and Hot Potatoes* and *The Spirit of the Soil,* and more than fifty scholarly articles on ethical and policy issues in agriculture, agrarian values, biotechnology, and agricultural research. Thompson is past president of the Agriculture, Food and Human Values Society.

Suzanne Vaupel is an agricultural economist and attorney practicing in Sacramento, California. She has written extensively on agricultural labor and organic agriculture. Before opening her law practice she was visiting agricultural economist at the University of California, Davis; staff counsel for the State of California; and director of the consulting firm Vaupel Associates. As recipient of a Charles A. Dana fellowship, she studied labor conditions in newly industrialized countries at the International Labour Organisation in Geneva.

Editor

William Lockeretz is professor at the School of Nutrition Science and Policy, Tufts University, where he has been on the faculty since 1981. He is editor of the *American Journal of Alternative Agriculture,* which he helped establish in 1984. Lockeretz is coauthor of *Agricultural Research Alternatives* and has edited five books on agricultural subjects, most recently the proceedings of the conference "Environmental Enhancement through Agriculture." His research has been concerned with organic farming and with agriculture near metropolitan areas. Twice he served on committees of the Board on Agriculture of the National Research Council. He is a fellow of the American Association for the Advancement of Science.

Index

ISBN 0-8138-2044-8